Time and Transformation in Architecture

At the Interface/Probing the Boundaries

VOLUME 100

The titles published in this series are listed at *brill.com/aipb*

Time and Transformation in Architecture

Edited by

Tuuli Lähdesmäki

BRILL
RODOPI

LEIDEN | BOSTON

Cover illustration: "Atomium in Brussels" by Tuuli Lähdesmäki. Copyright with the editor.

Library of Congress Cataloging-in-Publication Data

Names: Lähdesmäki, Tuuli, editor.
Title: Time and transformation in architecture / edited by Tuuli Lähdesmäki.
Description: Leiden ; Boston : Brill, 2018. | Series: At the Interface/Probing the Boundaries ; volume 100 | Includes bibliographical references and index.
Identifiers: LCCN 2018023874 (print) | LCCN 2018024340 (ebook) | ISBN 9789004376793 (e-book) | ISBN 9789004366404 (pbk. : alk. paper)
Subjects: LCSH: Architecture--Philosophy. | Time perception.
Classification: LCC NA2500 (ebook) | LCC NA2500 .T54 2018 (print) | DDC 720.1/08--dc23
LC record available at https://lccn.loc.gov/2018023874

Typeface for the Latin, Greek, and Cyrillic scripts: "Brill". See and download: brill.com/brill-typeface.

ISSN 1570-7113
ISBN 978-90-04-36640-4 (paperback)
ISBN 978-90-04-37679-3 (e-book)

This book is printed on acid-free paper and produced in a sustainable manner.

Contents

PART 3
Spirituality and Decay in Architecture

Preface

Time and transformation form a natural and ineluctable bond—all transformations, changes, and alterations take place in relation to time and can be perceived due to their temporal nature. This volume explores time and transformation in architecture, in which changes and alterations may appear either suddenly or as slow *longue durée* processes. Transformation in architecture not only means changes in the materiality, construction, or planning of buildings—it also implies transformation of meanings and interpretations of architecture and transforming experiences in architectural space.

Although temporal transformation is inevitable, humans often long for permanence. In architecture, this longing materializes in monumental buildings that are expected to challenge the course of time and manifest continuity through eras and epochs. Diverse 'heritagization' practices and legislative acts that seek to preserve architecture and urban spaces concretize communities' need to prevent or retard transformation. Simultaneously, preservation and 'heritagization' are acts that transform the 'original' meanings and functions of architecture—and are, thus, a part of its inevitable temporal change.

This book approaches the complex questions of time and transformation in architecture from multi- and interdisciplinary points of view. The chapters in this book are based on further development of conference papers presented at the 4th Global Conference on Space and Place: Exploring Critical Issues, held from the 9th to the 12th of September, 2013 in Mansfield College at the University of Oxford, Great Britain. The Space and Place conference series was organized by Inter-Disciplinary.Net with Senior Lecturer Dr. Matthew Melia from Kingston University, Great Britain, as its chair. The interdisciplinary conference gathered scholars from various fields and practitioners with divergent professional backgrounds.

Some of the conference papers presented at this conference explored space and place in architecture with a particular emphasis on time, temporality, and temporal change and their effects on buildings, housing projects, and uses of urban space. After the conference, I selected the most interesting papers on this topic and suggested editing a thematic volume based on these papers to the conference organizers. The papers were developed and elaborated into volume chapters in two phases. As a result of this development, this book includes three sections that approach the topic of time and transformation in architecture from different perspectives. The chapters communicate with each other not only within the sections, but also between them. This communication brings out how humans' spatial interaction, spatial practices, and spatial meaning-making processes may resemble each other despite divergent cultural and social contexts and geographical locations.

I want to thank all the contributors to this book for their thorough work in developing their chapters and for fluent communication during the finalizing phase of the volume. I also thank the Inter-Disciplinary.Net, the organizers of the Space and Place conference series, and Matthew Melia as the chair of these conferences for enabling fruitful, critical, interdisciplinary discussions on diverse spatial issues. I thank assistant editor Bram Oudenampsen from Brill for smooth cooperation in the publishing process. My contribution as the editor of this volume is a part of my research project 'Identity politics in Pécs, Tallinn, and Turku as European Capitals of Culture' (ID-ECC), funded by the Academy of Finland. I thank the Academy of Finland for funding the project funding and thereby enabling the editing of this book.

Tuuli Lähdesmäki
Jyväskylä, Finland
December 15, 2017

List of Illustrations

Figures

Tables

Notes on Contributors

Eiman Ahmed Elwidaa
is a PhD student at the Department of Urban and Rural Development at the Swedish University of Agricultural Sciences, Uppsala, Sweden. She is pursuing her PhD studies in the area of gender and housing. Her findings so far have proven that government provided low-income housing examples are smaller prototypes of formal housing whose roots can be traced in colonial urban planning.

June Jordaan
is a professional architect and lecturer at the Cape Peninsula University of Technology, Cape Town. Her research interests include architectural theory, phenomenology, hermeneutics and philosophies concerned with place. She has presented her research at international conferences in Oxford, Prague, Cape Town, Hong Kong, and Japan. In 2015 she completed her PhD, entitled *Constructing Place: Towards a Phenomenological Framework for Architecture in the 21st Century*, in which she articulates three phenomenological categories that enable the interpretation, conceptualization, and representation of place.

Joongsub Kim
PhD, AIA, AICP, a Professor at Lawrence Technological University (Southfield, Michigan, United States) directs its community design studio and Master of Urban Design Program. Kim focuses on public interest design, urbanism, and community development, and has received an ACSA/AIAS New Faculty Teaching Award; an ACSA Collaborative Practice Award citation; a Graham Foundation Advanced Studies Grant; an NCARB Integration of Practice and Education National Grant; an AIA Michigan President's Award; and grants from the National Endowment for the Arts and Knight Foundation. His work has been published in *Urban Design International*, *Journal of Urban Design*, *Places*, *Environment & Behavior*, and *Open House International*.

Tuuli Lähdesmäki
PhD, DSocSc, is an Academy Research Fellow and Adjunct Professor at the Department of Music, Art and Culture Studies, University of Jyväskylä (JYU), Finland. Lähdesmäki specializes in heritage, culture, and identity politics particularly in the European context. In her recent research project (ECC-ID) funded by the Academy of Finland, Lähdesmäki examined cultural politics,

identity-building, space, and urban development in the European Capital of Culture initiative. She is continuing research on these topics in her current research projects 'European Cultural Heritage in the Making: Politics, Affects and Agency' (EUCHE), funded by the Academy of Finland, and 'Legitimation of European Cultural Heritage and the Dynamics of Identity Politics in the EU' (EUROHERIT), funded by the European Research Council.

Assumpta Nnaggenda-Musana
PhD, PhL, is a Lecturer of Architecture in the Department of Architecture and Physical Planning at the College of Engineering, Design, Art and Technology, Makerere University, Kampala, Uganda. Her studies on formal and informal housing in Kampala, Uganda, have shown that urban sprawl was a creation of colonial planning systems. She has done several researches on informal settlements and low-income housing, finding ways of promoting houses that are user-friendly.

Sanja Rodeš
PhD, MArch, BArch, is a Lecturer in School of Architecture and Built Environment at Deakin University, Geelong, Australia. Her research interests are architectural history, theory and criticism of the late twentieth and twenty-first centuries, and she is particularly interested in the relationships between contemporary architecture, media and image.

Smaranda Spânu
is a practicing architect and has been conducting research in the fields of architectural theory, space and urban planning, restoration, and the built heritage since 2011. She has completed her PhD, focusing on the relationship of heterotopic spaces and the built heritage, in 2016 at Babes-Bolyai University in Cluj-Napoca, Romania, and her undergraduate studies at Faculty of Architecture and Urban Planning, Technical University of Cluj-Napoca, Romania. While working in a private practice, she has been engaged in urban planning projects as well as architectural design and restoration projects, in which she anchors her research rationale.

André van Graan
is an Associate Professor in the Department of Architectural Technology at the Cape Peninsula University of Technology, South Africa. His research focus has been on the impact of colonial modernism on indigenous human settlements, with particular reference to British colonial practice in the twentieth century.

CHAPTER 1

Introduction: Time and Spatial and Social Turns in Architectural Research

Tuuli Lähdesmäki

During the past four decades, the connections and interaction between space and social relations have raised considerable critical interest in academia. Among other much discussed scholarly 'turns' in human sciences, the spatial turn has influenced the understanding of reality as constructed and determined by complex spatial relations. The spatial turn has introduced to scholarly discussions concepts and metaphors borrowed, for example, from geography,[1] topography, and physics, and theoretical views developed in philosophy, cultural and urban geography, sociology, and anthropology. It has also participated in the blurring of disciplinary boundaries by increasing interest in the sociology of space in humanities and the spatiality of cultural practices in social sciences. Edward W. Soja has described the spatial turn as 'a response to a longstanding ontological and epistemological bias that privileges time over space in all the human sciences, including spatial disciplines like geography and architecture'.[2] In this view, Soja emphasizes temporality and spatiality as distinguished ontologies, of which the former has dominated scholarly discussions until the last decades of the 20th century. Simultaneously, the spatial turn in human sciences has led to a 'social turn' in disciplines that explore urban space, the built environment, and architecture. Due to the 'social turn', scholars in architectural history and theory have become increasingly interested in, for example, questions concerning class, power, gender, race, and ethnicity.[3]

1 According to Mike Crang and Nigel Thrift, the spatialized vocabulary of geography in other disciplines, such as cultural studies and more broadly in humanities, is often deployed in various theoretical discussions unrelated to or ignorant of geography as a discipline. Mike Crang and Nigel Thrift, *Thinking Space* (London: Routledge, 2000), xi.

2 Edward W. Soja, "Taking Space Personally", in *The Spatial Turn: Interdisciplinary Perspectives*, eds. Santa Arias and Barbara Warf (London: Taylor and Francis, 2008), 11.

3 Yael Allweil, "Beyond the Spatial Turn: Architectural History at the Intersection of the Social Sciences and Built Form", in *The Proceedings of Spaces of History / Histories of Space: Emerging Approaches to the Study of the Built Environment* (Berkeley: University of California, 2010), accessed October 5, 2017, https://escholarship.org/uc/item/9rt7c05f.

 | DOI 10.1163/9789004376793_002

Architecture and the built environment arise in the intersection of diverse cultural, social, societal, economic, and political relationships and their spatial dimensions. The spatial turn has brought these dimensions to the core of theoretical discussions on architecture. In architectural discourse, however, the concept of space as such did not have to be rediscovered; the idea of architectural creations as complex spatial constructs has been discussed in architectural theory since the end of the 19th century.[4] Nevertheless, the more multifaceted understanding of the intertwined nature of spatiality, social relations, power, and the built environment is more recent. The past four decades have seen in the research on architecture and the built environment a growing interest in the ideas of power and space as social products and particularly as producers of social reality. In these views, space is not only seen as a physical frame for human interaction but also as a performative catalyst of interaction and social relations. The idea of space as produced by interrelated social relations was introduced by Henry Lefebvre in his influential book *La production de l'espace*, published in 1974. In Lefebvre's theory, the physical environment is a complex junction of sensory, perceived, and materialized spaces intertwined with linguistic conceptualizations, symbolic representations, and subjective experiences, beliefs, and uses of space.[5] Lefebvre has discussed these different aspects of space as perceived (*le perçu*), conceived (*le conçu*), and lived (*le vécu*).[6]

In addition to Henri Lefebvre's and Edward W. Soja's views, the spatial turn in humanities and social sciences has broadly utilized Michel Foucault's notions on the ambiguous and productive nature of space. Foucault's concept of heterotopia has been particularly important for descriptions of a space that takes place and exists in social relations.[7] The loose and open nature of Foucault's concept has generated a variety of interpretations and applications. Kevin Hetherington interprets the meaning of the concept as constructed through the juxtaposition of the specific materiality of a place, its social practices, and particular events. According to him, heterotopias are, thus, simultaneously place-bound, socially-bound, and time-bound.[8]

4 Eve Blau, "Introduction," in *Repenser les limites: l'architecture à travers l'espace, le temps et les disciplines*, ed. INHA (Paris: INHA, 2005), 2; Łukasz Stanek, "Architecture as Space, Again? Notes on the 'Spatial Turn'", *SpecialeZ* 4 (2012): 49.

5 Henri Lefebvre, *The Production of Space*, trans. Donald Nicholson-Smith (Oxford: Blackwell, 1991), 38–41.

6 Ibid.

7 Michel Foucault, "Of Other Spaces: Utopias and Heterotopias," *Architecture /Mouvement/ Continuité*, translated by Jay Miskowiec, 5 (1984): 46–49.

8 Kevin Hetherington, *The Badlands of Modernity: Heterotopia and Social Ordering* (London: Routledge, 2002), 8.

The spatial turn has an ambivalent relation to time and aspects of temporality. On the one hand, the emphasis on spatiality has been interpreted—as Soja does—as a response to the privileged position of time in scholarly discussions in human sciences. On the other hand, the aspects of time and space have often been discussed as inseparably intertwined, as Foucault's notion of heterotopia exemplifies. Indeed, recent study on space has commonly focused on the relations between time and space. Space has been perceived as a historical construction bound as much to its past as to its present time. To stress the temporality of space, Doreen Massey has used the concept of time-space: spaces are not static entities but continuous processes that are influenced by their past.[9] In architectural history and theory, the interdependence of time and space has been a topic of discussion since the beginning of the 20th century. As Eve Blau notes, the views emphasizing space as relational and contingent and as both subjectively and objectively constituted over time and, therefore, perceived as being in a perpetual state of 'becoming', have been foundational not only for the discipline of architectural history, but also for early 20th century theorizing of modern architecture.[10] Modern architectural space, city planning, and art have been approached in these discussions, for example, as dynamic 'space-time', a concept used by Sigfried Giedion already in the 1940s.[11]

Architecture binds together imaginings and reality, materiality and intangible values. As much as architecture is about the physical environment, it is also about conceptualizations, narrations, discourses, symbolism, experiences, emotions, and effects. It embraces private and public, subjective and communal dimensions. Architecture is an event, as Bernard Tschumi has stated.[12] But it is also a business[13] and a product, which like any other commodity in capitalism, promotes the expansion of profit, the increase of production, and consumption.[14] The sphere of architectural discussions has extended and become multifaceted and flexible. The borders between architecture and non-architecture have loosened, and the conceptual categories between architecture and various

9 Doreen Massey, *For Space* (London: Sage, 2005), 177–180.

10 Blau, "Introduction," 2.

11 Siegfried Giedion, *Space, Time and Architecture: The Growth of a New Tradition* (Cambridge: Harvard University Press, 1941).

12 Bernard Tschumi, *Architecture and Disjunction* (Cambridge: MIT Press, 1994).

13 Umberto Eco, "Mass Appeal in Architecture," in *Rethinking Architecture: A Reader in Cultural Theory*, ed. Neil Leach (London: Routledge, 1997), 187–193.

14 Anna Klingmann, *Brandscapes. Architecture in the Experience Economy* (Cambridge: The MIT Press, 2007), 4.

other phenomena, such as art[15] and brand,[16] have become more elastic. This flexibility has also produced new ways of approaching architecture and conceptualizing phenomena related to it.[17]

The book at hand approaches architecture and the built environment from an interdisciplinary point of view, emphasized in its theoretical and empirical discussions on the dimensions of time, temporality, and transformation. Architecture is approached in the book in a broader social and societal framework than is often done in publications in the fields of art history, architectural history and theory, and cultural and urban studies. The book focuses on modern and contemporary architecture by bringing into the discussion the contradiction between the imaginings and reality of urban space and by exploring the material and conceptual borders of architecture. The fundamental questions penetrating the discussions of the book are: What is architecture? How do architectural ideas, ideals, and meanings emerge, develop, and transform? How is architecture manifested in relation to time, time-space, and the social dimensions it entails and produces? The book provides both multifaceted theoretical discussions on time and temporality in architecture and empirical case studies in which the explanatory power of these theories and conceptualizations are tested and explored.

Through divergent empirical cases, the book provides both local and global points of view on challenges of modern and contemporary architecture and the built environment as well as their relation to traditional architectural language. The geographical scope in the book includes several European cities (Tallinn in Estonia, Maribor in Slovenia, Marseille in France, and Turku in Finland) and regions (Transylvania in Romania), metropoles in the US (Detroit and New York) and China (Beijing), and African cities and towns (Cape Town in South Africa and Kampala and Jinja in Uganda). Moreover, the discussions in the book refer to architectural examples from several other cities around the globe. The focus in the empirical cases varies from skyscrapers to low-income housing, from public buildings to private apartments, from historical

15 Miwon Kwon, "For Hamburg: Public Art and Urban Identities," in *Public Art is Everywhere*, ed. Christian Phillip Müller (Hamburg: Kunstverein Hamburg and Kulturbehörde Hamburg, 1997), 95–109; Miwon Kwon, *One Place After Another: Site Specificity and Locational Identity* (Cambridge: MIT Press, 2002).

16 For example, Klingmann, *Brandscapes*.

17 One of the recent examples of creative conceptualizations of architecture and its relations to other spatial phenomena is Sylvia Lavin's concept of 'kissing', with which she describes the sensual charge and intimacy between architecture and new types of art, particularly multimedia installations that take place in and on the surfaces of buildings. Sylvia Lavin, *Kissing Architecture* (Princeton: Princeton University Press, 2011).

buildings to ongoing architectural projects, from secular to religious buildings, from protected built heritage to deteriorated and abandoned buildings, and from rural to urban space. Architecture in the case studies of the book has various aims: to create well-being, regenerate urban space, impress, control, commemorate, etc.

The book is composed of seven chapters that are organized into three thematic sections. The first section, 'Temporality and Changing Meanings of Urban Space', investigates the meanings of architecture in the intersection of political, economic, cultural, and social spheres. The section includes two chapters that discuss the transformation and meanings of urban space. Tuuli Lähdesmäki's chapter, entitled 'Temporal Architecture as a Means in Urban Regeneration', focuses on the ideas of temporality, movability, processuality, and transformation in contemporary architecture. These ideas contradict the Western architectural tradition, which has commonly embraced the ideas of durability and eternity, and has approached architectural constructions as materialized history and manifestations of the cultural ideas of past generations. However, all architectural creations have a certain life span, and in some cases the span is planned to be relatively short, and the durability of the creations is meant to be limited.

Lähdesmäki's chapter brings to the fore the close relation between architecture and economics in current cultural policy discourses and urban development and management talk. During recent decades, decision-makers and urban planners have sought to advance the economic growth of Western cities by promoting diverse cultural projects and initiatives aiming to develop the city space. These projects and initiatives have commonly included construction of new buildings, squares, parks, and other urban public spaces and recreational areas with new urban fittings. Scholars have discussed and conceptualized these aims with the concepts of urban revitalization and urban regeneration.[18] The aims of urban revitalization and regeneration are also intertwined with attempts to produce experiences. Indeed, urban planners and marketing experts have increasingly focused on the development of experiences in cities in order to foster consumption.

Thus, the current aims of urban planning reflect the ideas connected to the concept of experience economy. The concept—introduced by Joseph B. Pine and James Gilmore already in 1989—suggests that production of economic

---

18 Robert Beauregard and Briavel Holcomb, *Revitalizing Cities* (Washington, DC: Association of American Geographers, 1981); Sean Zielenbach, *The Art of Revitalization. Improving Condition of Distressed Inner-City Neighborhoods* (London: Garland Publishing, 2000); Robert Cowan, *The Dictionary of Urbanism* (Tisbury: Streetwise Press, 2005).

values relies more and more on intangible factors and above all on creating experiences.[19] Over the past 20 years, designing and building new spaces for the consumption of experiences has become the basis of many urban regeneration programs. Urban planners have tried to create places that succeed as venues for diverse consumable and memorable events.[20] Architecture has a significant role in experience economy; it acts as a catalyst for new experiences and perceptions and functions as a physical frame for consumption.[21] Scholars have emphasized the close connections between architecture and economy by indicating how architecture is integral to branding and vice versa. 'Brandscapes', physical sites demarcated by brands, have become key elements that link together identity, culture, and place in contemporary cities.[22]

In her chapter, Lähdesmäki discusses how current attempts at urban revitalization and regeneration are promoted not only at the local level, but also at the European level by the European Union (EU). The EU's urban initiative, the European Capital of Culture, encourages European cities to promote cultural regeneration as a part of their development. The limited length of the designation (one year) has led the designated cities to implement and host various temporary architectural projects in which the city space is intervened through various tactics to reach diverse cultural, social, political, and economic goals. Lähdesmäki introduces four temporary architectural projects from four recent European Capitals of Culture; Tallinn, Turku, Maribor, and Marseille. The chosen cases indicate how the idea of temporality opens a variety of new possibilities for rethinking and practicing architecture.

Sanja Rodeš's chapter, entitled 'After the Event: Considering Media Images and Skyscraper Architecture at the Turn of the 21st Century', discusses an interactive relationship between architecture, image, and media. The chapter focuses on the events of 9/11 in New York and the impact of the destruction of the Twin Towers to the meanings and reception of two recent skyscraper projects. These projects are the new World Trade Center (WTC) in New York (2006–present), initially designed by architect Daniel Libeskind, and the China Central Television Headquarters in Beijing, China (opened in 2012), designed by the Office for Metropolitan Architecture led by architect Rem Koolhaas.

19 Joseph B. Pine and James H. Gilmore, *The Experience Economy* (Boston: Harvard Business School Press, 1989).

20 Greg Richards and Robert Palmer, *Eventful Cities: Cultural Management and Urban Revitalisation* (London: Routledge, 2010), 74.

21 Klingmann, *Brandscapes*, 2.

22 Ibid., 2–3, 83.

The 2001 terrorist attack gave the architecture and place of the Twin Towers a special symbolic and emotive meaning, which was reinforced by the ensuing global media spectacle. This symbolic and emotive meaning has been widely discussed—and at the same time maintained and strengthened—in various disciplines in academia, such as media studies, visual culture studies, and architectural theory. Skyscrapers have commonly been interpreted as signs of an energetic economy.[23] They are also crucial elements in creating recognizable cityscapes and city brands. The Twin Towers were a significant part of the visual symbolism of the New York skyline, since they were the tallest buildings in the city. The iconic towers that embraced economic, political, and touristic meanings were reproduced and circulated in billions of images.

Rodeš approaches the new WTC project by utilizing the idea of 'limits'—boundaries that define architecture—to understand and explain the meanings that 9/11 created for the project. She also approaches 9/11 as an iconotype, an image that through repetition stands out in the image flow, halts the incessant circulation, and draws fixity to itself.[24] The idea of 9/11 as an iconotype reflects the role of images and visuality in contemporary culture. The intensified visual culture forms the prerequisite for the emergence of iconotypes. Moreover, Rodeš explains the relationship between 9/11 and the WTC architecture with the concept of iconomy, 'image economy'.[25] According to this point of view, the images of spectacular architecture, events, contemporary art, etc., gain exchange value in an economy due to the effects of mediatization and spectacularization. The destruction of the Twin Towers and the construction of the new WTC project have transformed the meanings of the site of the 9/11 attack. The site of destruction has been transformed into a site of reconstruction. The meanings of tourism and constant flows of people have been added to the site that was filled with the meanings of fear, anxiety, mourning, and trauma.

Besides iconomy, Rodeš utilized in her chapter the concept of translation, which relies on the existence of an iconomy and functions as a means to explore and explain image flows. Architecture is commonly 'translated' into images that are further translated into other architectural images. The circulation of architectural images simultaneously produces translations of architectural concepts and ideas. In her chapter, Rodeš explores how the 9/11 iconotype is translated into the architectural concept of the China Central Television Headquarters. Rem Koolhaas has himself correlated and compared this skyscraper

23 See for example, Terry E. Smith, *The Architecture of Aftermath* (Chicago: University of Chicago Press, 2006).

24 Smith, *The Architecture of Aftermath*, 22.

25 Ibid., 5–8.

with the Twin Towers. The chapter brings out how images of architecture and an urban environment are often used as a representational tool for communicating particular public messages.

Architecture is both about imaginings and reality. It brings to the fore certain ideas and ideals that its creators seek to communicate through it, but it also reflects meanings that the transforming cultural, social, societal, and political contexts and time produce for it. The second section in the book is entitled, 'Architectural Imaginings and Colonial and Post-Colonial Realities', and it includes three chapters that discuss urban planning and housing policies and practices in Africa. The section brings to the fore challenges that the colonial past poses for housing ideas and architectural meanings. The section starts with André van Graan's chapter, 'Modernism as a Mechanism of Power and Control in Colonial Contexts: The Project of Modernity in Cape Town, South Africa'. The modernist movement is one of the turning points in Western architecture; it has had an enormous impact on the development of cities, industrial production, housing, living environments, and the 20th and 21th centuries' way of everyday life. The foundation of the modernist movement was in the ideas and ideals of improving people's social and physical well-being.[26] In many European countries, the ideas of modern architecture were used to produce welfare in broader societal terms—to develop the welfare state.[27]

Modernist architectural ideas spread from the Western world to its colonies, where the ideals of well-being and the interests in improving the living environment of the ordinary citizen, or the 'little man in the street', to use the expression of modernist architect Alvar Aalto,[28] were often dismissed and forgotten. As the chapter by van Graan indicates, in late colonial countries, such as South Africa, modern architecture became primarily a mechanism for improving production, restructuring urban areas into racially and physically segregated areas, and dealing with issues related to health and slum clearance.

André van Graan emphasizes that the discourse on architectural modernism, particularly in the colonial context, is strongly grounded in power relations, mechanisms of control, and the terrains of contestation and negotiation of difference. He explains the nature of architecture as a mechanism of social engineering by including in the discussion Foucault's and Lefebvre's theories

---

26 Peter Hall, *Cities of Tomorrow: An Intellectual History of Urban Planning and Design Since 1880*, 4. edition (Chichester: Wiley-Blackwell, 2014).

27 Mark Swenarton, Tom Avermaete and Dirk van den Heuvel, eds., *Architecture and the Welfare State* (New York, Routledge, 2015).

28 Sarah Menin and Flora Samuel, *Nature and Space: Aalto and Le Corbusier* (New York: Routledge, 2002).

on the significance of space in social discourse and in the construction of power hierarchies. Van Graan suggests that the catalyst behind the modernist movement in architecture in general was not only in the attempts to realize social reform, but also to react to the social and political situation in the Western world. The ideas and ideals of modern architecture included a possibility to use power against the threats produced by the political and social changes in Europe. As van Graan writes:

> Although social reform is the often cited reason, one wonders whether the aftermath of the First World War and the Bolshevik revolution in Russia was not feared as a threat to established order and whether they led to new Foucaultian methods of social control and surveillance to avert any possible threat. Modernism became the mechanism for visibly demonstrating change: the visualisation of political will.[29]

Similarly, in the late colonial countries modernism created spaces that rendered the 'Colonial Other' visible and controllable. The concept of the colonial 'Other' forms an important part of the postcolonial theory and explains how the colonized were dealt with as outsiders. In the early twentieth century South Africa, the division of people into 'us', who were white, and the 'Others', who were not, multiplied into much more heterogeneous clusters of subdivisions, taking the racist discourse to a new, intricately nuanced level. Van Graan notes how the outcome of this differentiation and the fixing of 'Otherness' to ethnic groups was used to map out the differing planning approaches that the City Council of Cape Town applied to different groups in its housing policies. Modernist architecture became the visual politics of the state as it suited its racist political agenda.

The second chapter in the section continues the discussion on spatial meanings in South Africa. June Jordaan's chapter, 'Architectural Agency and "Place-Making" in a Transformative Post-Apartheid South African Landscape', focuses on post-apartheid place-making practices which participate in the construction of a new South African identity. Jordaan emphasizes how in South Africa 'place' is an ideologically loaded word that has often been referred to as a 'healing instrument' through which the discrimination of apartheid and its spaces can be redressed. In order to engage with the creation of a post-apartheid South African identity, several architects have sought to reinterpret

29 André van Graan, "Modernism as a Mechanism of Power and Control in Colonial Contexts: The Project of Modernity in Cape Town, South Africa," in this volume.

the meaning and significance of particular locales. These spatial and architectural reinterpretations include various place-making strategies.

In her chapter, Jordaan introduces two opposing theories of place-making. 'Finite, bounded, and differentiated space' is characterized by features Jordaan describes with concepts such as 'manicured', 'disneyfied', 'controlled', 'big narratives', and 'fixed meanings', while 'infinite, open, and fluid space' allows, for example, 'spontaneity', 'chaos', 'change', 'small narratives', and 'ambivalent meanings'. The distinction between these two place-making strategies is related to time: the former denies the contingencies that are brought about by time, while the latter inclines to environments that facilitate time and the contingencies it implies. As Jordaan indicates, recent South African architectural projects have followed one of these strategies, but can also be identified as 'post-apartheid hybrids'.

Jordaan deepens the discussion on architecture and spatiality by exploring how body, space, and time are intertwined. Drawing from the ideas of Lefebvre and Soja,[30] she identifies three spatial dimensions of the life world: material, lived, and mental. This spatial triad is applied in an investigation of the meanings of a special South African place, the Cape Town Railway Station. In its history, the station has been subject to two major architectural place-making interventions. The earlier occurred in a late colonial and apartheid context in which South African modernist architecture was appropriated as 'a mechanism of power and control', as André van Graan describes it in his chapter, and the latter, in a democratic and transformative post-apartheid context that gave the oppressed members of society access to and free movement in the station. The newest interventions in the station were designed and timed for South Africa's hosting of the 2010 Soccer World Cup.

Jordaan notes how a multidimensional approach to place and space, which incorporates time, is useful to explicate various connotations and meanings of architecture which the architect him/herself cannot influence. She emphasizes that places should be considered as phenomena that affect 'life worlds' in various ways and on various scales: places are tied up with the way people use them and the meanings they attach to them. We do not live in separate material and mental worlds; these dimensions of 'life worlds' are fully intertwined.

The third chapter in the section is entitled 'Women as Retrofits in Modernist Low-Income Housing'. Its authors, Assumpta Nnaggenda-Musana and Eiman Ahmed Elwidaa, discuss Uganda's housing policies from the point of view of gender. Gendered spaces and spatial politics have been broadly discussed in

30 Lefebvre, *The Production of Space*; also see Edward Soja, *Thirdspace: Journeys to Los Angeles and Other Real-and-Imagined Places* (Oxford: Blackwell, 1996).

Western multidisciplinary feminist scholarship.[31] The empirical focus of these discussions has, however, often been limited to the Western geographical locus. There is a significant practical need for research utilizing the point of view of gender beyond its common context in Western scholarly discussion, as this chapter indicates.

Similar to the South African context discussed in the previous chapters, colonial urban planning in Uganda was based on an ideology of economic and racial segregation. Racial segregation policies divided the urban housing areas into British, Asian, and African neighborhoods, enabling the British to monitor and control the 'lower' populations. The chapters by van Graan and Nnaggenda-Musana and Elwidaa bring to the fore how colonial government policies promoted construction of houses with few rooms and only basic facilities to accommodate African short-term male laborers. Nnaggenda-Musana and Elwidaa note that in the post-colonial period, urban planning in the major cities of Uganda was—and still is—guided by colonial planning principles and building regulations, since no major modifications have been made to these principles after the colonial period.

Nnaggenda-Musana and Elwidaa examine in particular the current low-income housing policies of the Ugandan government. Regardless of new ideas and points of view that encourage practical, flexible, and gender-sensitive design of living environments, the size of the low-income houses in the government's housing programs is similar to the temporary housing of male laborers in the African neighborhoods of the colonial era. How should the post-colonial housing policies take into account the needs of both genders in order to make women's lives easier for collaborative and productive income-generating activities, while enabling general household activities? Gender is an important but complex point of view in housing policies, since gender is a profoundly intersectional social category. Various other social categories and subject positions, such as race and class, influence gender relations. As Nnaggenda-Musana and Elwidaa note, the needs and concerns of the women who belong to a particular group might differ greatly from those of another group.

The empirical focus in Nnaggenda-Musana's and Elwidaa's chapter is on the MWHS: Masese Women's Housing Scheme in Jinja—one of the settlement upgrading schemes implemented in Uganda in the late eighties. The scheme had women as the main beneficiaries. The investigation indicates how the women in the settlement have modified their neighborhood, tamed and developed it

31 See for example, Doreen Massey, *Space, Place, and Gender* (Minneapolis: University of Minnesota Press, 1994); Iain Borden, Barbara Penner and Jane Rendell, eds., *Gender Space Architecture: An Interdisciplinary Introduction* (*Architext*) (New York: Routledge, 1999).

to suit their ways of living better, and turned the neighborhood into a vivid and emancipating living environment.

The last section in the book is entitled, 'Spirituality and Decay in Architecture'. It exemplifies the temporality of architecture and discusses how changes in social, societal, political, economic, and cultural conditions and values influence the life span of buildings. The first chapter in the section, entitled 'The Heterotopic Nature of the Built Heritage. The Sacred Wooden Architecture of Transylvania and Its Practices', is written by Smaranda Spânu. The chapter focuses on old Transylvanian rural churches in Romania and the transformation of meanings that takes place when an ecclesiastical sacred space is protected and labeled as heritage. Through a process of 'heritagization', buildings which used to be the centers of rural villages have become isolated enclaves, removed not only from their original use but also from the natural flow of time, turning them into 'other' to their original context. Preservation of these churches as heritage seeks to tackle the passing of time; they are turned into a representation of a 'slice of time', living in their own 'parallel time' but simultaneously showcasing an idea of 'original time'. The materiality and physicality of the churches are sought to be preserved (thus, stopping their time flow) and in some cases restored (thus, reversing their time flow), as Spânu writes. The idea of a heritage object that is preserved and transferred as unchanged to future generations is driven by a struggle to suspend the flow of time and an impossible aim to freeze the materiality of objects detached from time.

Spânu brings to the discussion the hybrid spatial meanings of protected old churches, describing them as sacred ritual spaces, desacralized public spaces, mirrored spaces (since a new, bigger, and more robust church was often built next to the old one), and contested spaces (since their original use has been ceased). In general, heritage objects embrace a variety of values that change and whose relations change over the course of time.[32] Spânu emphasizes that the meanings and values are not, however, simply erased or replaced when the contexts of heritage change; they are often juxtaposed and remain imprinted within the physical form of the object—particularly when the heritage object is relatively immovable, like a building. When the social, cultural, historical, economic, and technological contexts are altered, some layers of values and meanings of heritage objects seem to be reinforced and accumulate, while some others seem to disappear. The layers that seem to disappear, however,

32 See for example, Alois Riegle, "The Modern Cult of Monuments: Its Character and Its Origin," *Oppositions* 25, Fall (1982): 21–50; Françoise Choay, *The Invention of the Historic Monument*, trans. Lauren M. O'Connell (Cambridge: Cambridge University Press, 2001).

only lose their visibility, although traces of the layers remain in the materiality of the object. Contemporary heritage policies include the idea of acknowledging all the layers of meanings and the varying values of heritage objects and their complex and even contradictory relationships.

Spânu finds Foucault's concept of heterotopia useful in discussing the various layers of meanings of old Transylvanian wooden churches. According to her, 'a built object can harbor a heterotopic potential within its material and spatial characteristics, since these are the direct outcome of specific social orderings.'[33] According to Spânu, the attempts to retain and preserve the old churches have added to them a new heterotopic layer of meanings, which has had both positive and negative outcomes. On the one hand, it has led to an increase in awareness of their various values and conservation of their physical form. On the other hand, the preservation has led to isolation and physical deterioration, since the heritage status has prohibited their physical alteration.

The section ends with the chapter entitled 'Understanding Spirituality in Disabled Places: Focusing on Urban Ruins and Urban Decay' by Joongsub Kim. The chapter discusses the experiences of spirituality that people may find in decayed buildings and an urban environment. The empirical focus of the chapter is on Detroit and the urban decay caused by the economic depression and the post-industrial structural changes that the city has faced after the recession in the car industry. The urban decay of the city has, however, recently raised intellectual, commercial, and touristic interest. In addition, the ruins arouse spiritual interest and spiritual experiences.

The chapter indicates how the urban ruins generate various emotional responses: a sense of hope, despair, disappointment, sadness, even anger. They also produce various affective meanings, such as empowerment; a sense of community, collective responsibility, and permanence; sacredness and religious experience; senses of place and attachment; ethics; and soulfulness. A common thread that penetrates the experiences of urban ruins is spirituality. As a result of participatory action research with focus groups, Kim suggests that the spirituality of urban decay can be defined in four ways:

> Some places in urban decay can be spiritual because they inspire people to do good things for others and the community (catalytic); places in urban decay can be spiritual because they are consoling (therapeutic); places in urban decay can be spiritual because they help connect individuals

33 Smaranda Spânu, "The Heterotopic Nature of the Built Heritage. The Sacred Wooden Architecture of Transylvania and Its Practices," in this volume.

> to their inner selves (reflective); and places in urban decay can be spiritual because they connect people in many different ways (engaging).[34]

As Kim's chapter indicates, urban decay does not only have negative meanings, it can also include positive connotations and evoke a sense of spirituality. This point of view on ruins is beneficial to cities such as Detroit that have suffered from extensive economic depression, deterioration and negligence of urban space, and related chronic urban problems.

Space and spatiality are fundamental elements in diverse social and societal domains and areas of everyday life; human activity cannot be separated from space and spatial relations. The book at hand demonstrates this by paying special attention to time, temporality, and the temporal transformation of architecture, the built environment, and the meanings of urban space—and their relations to human's experiences, behavior, and practices. These topics are approached in the book through diverse methods with a particular emphasis on hermeneutic, phenomenological, ethnographic, and anthropological approaches in exploring spatial experiences, identities, and cultural narratives. The multi- and interdisciplinarity of these discussions is ensured by the manifold scholarly expertise of the authors, who represent the disciplines of architecture, art history, cultural studies, development studies, heritage studies, and urban planning.

Bibliography

Allweil, Yael. "Beyond the Spatial Turn: Architectural History at the Intersection of the Social Sciences and Built Form." In *The Proceedings of Spaces of History / Histories of Space: Emerging Approaches to the Study of the Built Environment*. Berkeley: University of California, 2010. Accessed October 5, 2017. https://escholarship.org/uc/item/9rt7c05f.

Beauregard, Robert and Briavel Holcomb. *Revitalizing Cities*. Washington, DC: Association of American Geographers, 1981.

Blau, Eve. "Introduction." In *Repenser les limites: l'architecture à travers l'espace, le temps et les disciplines*, edited by INHA, 2–11. Paris: INHA, 2005.

Borden, Iain, Barbara Penner and Jane Rendell, eds. *Gender Space Architecture: An Interdisciplinary Introduction (Architext)*. New York: Routledge, 1999.

34 Joongsub Kim, "Understanding Spirituality in Disabled Places: Focusing on Urban Ruins and Decay," in this volume.

Choay, Françoise. *The Invention of the Historic Monument*. Translated by Lauren M. O'Connell. Cambridge: Cambridge University Press, 2001.

Cowan, Robert. *The Dictionary of Urbanism*. Tisbury: Streetwise Press, 2005.

Crang, Mike and Nigel Thrift. *Thinking Space*. London: Routledge, 2000.

Eco, Umberto. "Mass Appeal in Architecture." In *Rethinking Architecture: A Reader in Cultural Theory*, edited by Neil Leach, 187–193. London: Routledge, 1997.

Foucault, Michel. "Of Other Spaces: Utopias and Heterotopias." Translated by Jay Miskowiec. *Architecture /Mouvement/ Continuité* 5 (1984): 46–49.

Giedion, Siegfried. *Space, Time and Architecture: The Growth of a New Tradition*. Cambridge: Harvard University Press, 1941.

Hall, Peter. *Cities of Tomorrow: An Intellectual History of Urban Planning and Design Since 1880*, 4. edition. Chichester: Wiley-Blackwell, 2014.

Hetherington, Kevin. *The Badlands of Modernity: Heterotopia and Social Ordering*. London: Routledge, 2002.

Klingmann, Anna. *Brandscapes. Architecture in the Experience Economy*. Cambridge: The MIT Press, 2007.

Kwon, Miwon. "For Hamburg: Public Art and Urban Identities." In *Public Art is Everywhere*, edited by Christian Phillip Müller, 95–109. Hamburg: Kunstverein Hamburg and Kulturbehörde Hamburg, 1997.

Kwon, Miwon. *One Place After Another: Site Specificity and Locational Identity*. Cambridge: MIT Press, 2002.

Lavin, Sylvia. *Kissing Architecture*. Princeton: Princeton University Press, 2011.

Lefebvre, Henri. *The Production of Space*. Translated by Donald Nicholson-Smith. Oxford: Blackwell, 1991.

Massey, Doreen. *Space, Place, and Gender*. Minneapolis: University of Minnesota Press, 1994.

Massey, Doreen. *For Space*. London: Sage, 2005.

Menin, Sarah and Flora Samuel. *Nature and Space: Aalto and Le Corbusier*. London: Routledge, 2002.

Pine, Joseph B. and James H. Gilmore. *The Experience Economy*. Boston: Harvard Business School Press, 1989.

Richards, Greg and Robert Palmer. *Eventful Cities: Cultural Management and Urban Revitalisation*. London: Routledge, 2010.

Riegl, Alois. "The Modern Cult of Monuments: Its Character and Its Origin." *Oppositions* 25, Fall (1984): 21–50.

Smith, Terry E. *The Architecture of Aftermath*. Chicago: University of Chicago Press, 2006.

Soja, Edward W. *Thirdspace: Journeys to Los Angeles and Other Real-and-Imagined Places*. Oxford: Blackwell, 1996.

Soja, Edward W. "Taking Space Personally." *The Spatial Turn: Interdisciplinary Perspectives*, edited by Santa Arias and Barbara Warf, 11–35. London: Taylor and Francis, 2008.

Stanek, Łukasz. "Architecture as Space, Again? Notes on the 'Spatial Turn'." *SpecialeZ* 4 (2012): 48–53.

Swenarton, Mark, Tom Avermaete and Dirk van den Heuvel, eds. *Architecture and the Welfare State*. New York: Routledge, 2015.

Tschumi, Bernard. *Architecture and Disjunction*. Cambridge: MIT Press, 1994.

Zielenbach, Sean. *The Art of Revitalization. Improving Condition of Distressed Inner-City Neighborhoods*. London: Garland Publishing, 2000.

PART 1

Temporality and Changing Meanings of Urban Space

CHAPTER 2

Temporary Architecture as a Means in Urban Regeneration

Tuuli Lähdesmäki

Architecture and Time

Architecture is both a product and a process that combines not only esthetics and functionality, but also a broad variety of cultural, social, societal, and political meanings. It binds together materiality and immaterial and intangible values. Since antiquity and the influential architectural theories of Vitruvius (c. 80–70 BC to c. 15 BC), the ideals of Western architecture have comprised an aim for durability, or *firmitas* in Vitruvius's terms.[1] Indeed, the ideas of temporality and movability do not belong to the traditional notions of Western architecture. On the contrary, these notions often carry within them the idea of eternity: architecture is seen as something that is transmitted from one generation to another and with this continuity it transforms into the history of the community. As materialized history, architecture is often considered to comprise the values of the time of its construction and to be able to transmit them to the present. In the disciplines of art history, cultural history, and heritage studies, architecture is often approached and explored as cultural heritage—as artifacts that manifest and scrutinize the cultural ideas of the past generations. Due to this heritage function and value, preservation and conservation of architecture are considered important not only within the views of these disciplines, but also more broadly in Western societies. These traditional notions of Western architecture are manifested in the narration of Western architectural history. It is commonly told as a continuation of impressive creations of famous architects and a series of iconic landmarks and cultural heritage sites in major cities of the Western world.

In reality, architecture is never eternal, but rather transforming and temporal. The construction of the so-called masterpieces of Western architectural history has also often been processual. It may have taken centuries to build them and, thus, their architecture is a combination of various historical layers. The form, style, spatial structure, function, and use of these buildings may

1 Vitruvius, *The Ten Books on Architecture*, trans. Morris Hicky Morgan (Cambridge: Harvard University Press, 1914).

 | DOI 10.1163/9789004376793_003

have changed several times during their history. Gothic churches are interesting examples of the processual dimensions and transforming characteristics in architecture. Many Gothic churches were built between the 13th and 16th centuries, but in some cases their construction was not finished until the 19th century. For example, construction of the Gothic Cologne Cathedral was started in 1248 and was finally completed in 1880. In Barcelona, the Art Nouveau church Sagrada Família by Antoni Gaudí was commenced in 1882 and is still under construction. The iconic cathedrals and churches in European cities are also good examples of the transforming functions of architecture. Buildings that used to serve mainly religions purposes have been transformed into touristic spaces, museumized and heritagized sites, and visual icons commonly utilized for marketing, place promotion, and city branding, and, thus, changing the meanings of architecture.[2]

All architectural creations have a certain life span, and in some cases the span may be planned to be relatively short. Many buildings are intentionally constructed with methods and materials that have limited durability. Buildings may have been planned as temporary due to a sudden and temporal need to house a large group of people, such as immigrants, refugees, or soldiers. In addition, several venues for mega-events, such as sport games, have been constructed to be used only during the events. For example, in the London Olympic Games in 2012, several sports halls and arenas were constructed with materials and techniques that enabled downsizing, dismantling, and recycling of the buildings and/or moving and rebuilding them in another place.

Western architectural history includes a certain architectural genre—pavilions and venues of international exhibitions and world fairs—in which the architectural creations are intentionally planned to exist only temporarily while at the same time being profoundly outstanding. Many of these pavilions are today known only from photographs and drawings, but some of them have turned out to be much longer-lasting than was originally planned. In some cases these pavilions and exhibition venues, such as the Eiffel Tower (1889) in Paris or the Atomium (figure 2.1) in Brussels (1958), have become iconic landmarks and tourist attractions and acquired a heritage value, and they function as manifestations of ideas and world views of the time of their construction.

The idea of temporality has been recently much discussed and explored in contemporary architecture. Recent practices of urban regeneration have utilized the possibility of creating temporary architectural spaces to enliven

2 Dean MacCannell, *The Tourist. A New Theory of the Leisure Class* (New York: Schocken Books, 1989), 43.

FIGURE 2.1
Atomium, designed by André Waterkeyn, André Polak, and Jean Polak, 1958, Brussels, Belgium
PHOTO: TUULI LÄHDESMÄKI.

less-used or abandoned sites, to offer the local people venues for various communal activities, and to entice both the citizens and visitors in the cities to spend time in them. The idea of temporality opens a variety of new possibilities for thinking about and practicing architecture.

Since 2000, the European Union (EU) has become more and more interested in urban development as a tool for influencing European cities. One of the EU's longest running and perhaps most well-known and influential cultural initiatives—the annual designation of the European Capital of Culture (ECOC)—encourages European cities to promote local and economic development through diverse cultural activities and urban projects. The initiative was launched in 1985 and run as a European-level intergovernmental scheme under the name European City of Culture until 1999, when it was transformed into a Community action of the European Parliament and Council and renamed European Capital of Culture. Many of the designated cities have invested in large-scale infrastructure and construction projects before, during, and/or after their ECOC year. Moreover, the limited length of the ECOC designation (one year) has led the cities to implement and host various small-scale temporary architectural projects in which city spaces are altered through various tactics to achieve diverse cultural, social, political, and economic goals.

In this chapter, I examine the diversity of tactics and goals of temporary architectural projects in four ECOCs—Turku (Finland), Tallinn (Estonia), Maribor (Slovenia), and Marseille (France)—and discuss how temporary architectural intervention into an urban space is intended to work as a means of urban regeneration. The research material of the study consists of diverse planning and promotional materials of the architectural project cases, their observation *in situ*, and discussions about the projects with various local people and cultural actors during my field research and visits to the cities. This rich and manifold research material is approached with data-driven inductive logic of inquiry that seeks to understand temporary architecture in the context of recent theoretical discussions and conceptualizations of cultural policy research, urban studies, and culture studies.

The chapter starts from a brief theoretical discussion on the role of architecture in urban regeneration followed by a discussion on the ECOC initiative and the selected temporary architectural projects and the tactics of temporality and the goals of regeneration included in them. The article ends with a discussion on these tactics and goals in the contexts of recent urban planning theories and general conclusions about the study.

The Role of Architecture in Urban Regeneration

The intertwined emphasis on economic and cultural development and the interconnections between the spheres of culture and economy characterize current discourses on cultural policy in the Western world.[3] The spheres of culture and economy form an inter-discursive domain in which culture is commonly transformed into an economic keyword.[4] The concepts of creative economy and cultural economy—which have often been referred to in cultural policy discourses and urban development and management talk during the past two decades—are based on the idea of a correlation between cultural and economic development: strengthening one of them is considered to advance

3 Lluís Bonet and Emmanuel Négrier, "The Ends(s) of National Cultures? Cultural Policy in the Face of Diversity," *International Journal of Cultural Policy* 17, no. 5 (2011): 574–589.

4 See for example, Hans Erik Näss, "The Ambiguities of Intercultural Dialogue: Critical Perspectives on the European Union's New Agenda for Culture," *Journal of Intercultural Communication* 23 (2010).

the other, as well. In addition, the economic dimension of culture has often been intertwined with urbanity. Cities have been considered drivers of economic growth, with urban space as a factor supporting that function.[5] Thus, in cultural policy discourses and urban development and management talk, the ideas of urban space, culture, creativity, and economy form an interrelated and inter-determined sphere of meanings.[6]

During recent decades, decision-makers and urban planners have sought to advance the economic growth of Western cities by promoting diverse cultural projects and development of the city space. These aims have been discussed and conceptualized within academia with the concepts of urban revitalization and urban regeneration. What kinds of interventions do the concepts of regeneration and revitalization eventually refer to? According to Robert Beauregard and Briavel Holcomb, urban revitalization means putting 'new life' into cities and upgrading areas for 'higher' social and economic uses.[7] Sean Zielenbach (2000) has identified two distinct approaches to revitalization: individual-based approaches focus on people and improvement of their living conditions (such as preventing unemployment), while place-based approaches stress local economic development and raising real estate values (such as actions that produce gentrification).[8] The concepts of urban revitalization and urban regeneration are often used interchangeably. However, some authors, such as Robert Cowan, have distinguished the concepts by explaining urban regeneration to refer mainly to physical change and improvements of the environment, while revitalization is considered to refer to social improvements and other ways (such as cultural) of bringing a place to life.[9]

Culture may, however, have a crucial role in the physical transformation and development processes of urban space. Therefore, some scholars, like Graeme Evans, have examined the role of culture in urban regeneration in more detail. He has recognized three modes of incorporating cultural activity into the regeneration processes. He calls them: (a) culture-led regeneration (cultural activity as a catalyst and engine of regeneration); (b) cultural regeneration

5 Kaarin Taipale, *Cities for Sale. How Economic Globalization Transforms the Local Public Sphere* (Espoo: Helsinki University of Technology, 2009), 42.

6 Tuuli Lähdesmäki, "European Capital of Culture Designation as an Initiator of Urban Transformation in the Post-Socialist Countries," *European Planning Studies* 22, no. 3 (2014): 481–497.

7 Robert Beauregard and Briavel Holcomb, *Revitalizing Cities* (Washington, DC: Association of American Geographers, 1981).

8 Sean Zielenbach, *The Art of Revitalization. Improving Condition of Distressed Inner-City Neighborhoods* (London: Garland Publishing, 2000).

9 Robert Cowan, *The Dictionary of Urbanism* (Tisbury: Streetwise Press, 2005).

(cultural activity integrated into an area strategy alongside other activities in the environmental, social, and economic spheres); and (c) culture and regeneration (cultural activity is not fully integrated into the strategic development of the area).[10] On the one hand, decision-makers and urban planners seem to believe that physical transformation of urban space is necessary to bring about urban development and stimulating economic growth.[11] On the other hand, during the past couple of decades, culture-driven regeneration has occupied a pivotal position in contemporary urban policies.[12] The combined aims of physical transformation of urban space and various cultural interventions into it form a common strategy in today's urban planning.

A commonly used and concrete means for cultural regeneration in contemporary cities is architecture. The past couple of decades have witnessed a variety of large-scale urban projects through which cities have sought to tackle their declining industries and economic regression by investing in major cultural institutions, such as cultural centers, art museums, and opera houses, and by constructing for these institutions unique buildings meant to function as global landmarks.[13] Well-known examples of this kind of 'wow-architecture' range from the Sydney Opera House (1973) to the Bilbao Guggenheim Museum (1997), and from the Centre Georges Pompidou in Paris (1977) to the National Centre for the Performing Arts in Beijing (2007). The logic of 'wow-architecture' is entangled in its expected economic impact. World-famous architects have been invited to design outstanding and spectacular buildings in order to attract visitors and cultural actors, raise media attention, transform the city image, draw new investments to the city, etc. This kind of cultural regeneration strategy is extremely risky, though, because of the massive scale, expense, and often limited substantial focus of these buildings. The 'wow-architecture' and cultural infrastructure-based projects are also relatively inflexible strategies of cultural regeneration and image-making.[14] The concrete results of this kind of 'hard branding' of a city—based on the reputation and signature of global

10 Graeme L. Evans, "Measure for Measure: Evaluating the Evidence of Culture's Contribution to Regeneration," *Urban Studies* 42, no. 5–6 (2005): 1–25.

11 See for example, Teodoro Luque-Martinez, Salvador Del Barrio-García, José Ángel Ibáñes-Zapata, and Migual Ángel Rodriguez-Molina, "Modeling a City's Image: The Case of Granada," *Cities* 24, no. 5 (2007): 335–352.

12 Jana Temelová, "Urban Revitalization in Central and Inner Parts of (Post-socialist) Cities: Conditions and Consequences," in *Regenerating Urban Core*, ed. Tuomas Ilmavirta (Espoo: Helsinki University of Technology, 2009), 15.

13 Kai Horstmannshoff, *The Loop. Chicago Architecture and the Social Imaginary* (Bielefeld: Transcript, 2018).

14 Ronan Paddison, "City Marketing, Image Reconstruction and Urban Regeneration," *Urban Studies* 30, no. 2 (1993): 339–350.

architects—and its impact on urban regeneration may be difficult to prove.[15] The formula of constructing new spectacular cultural buildings to regenerate the city and its surrounding region and to transform the city's urban image may have been successful in some cities, but has failed at other sites and in different contexts. In the regeneration processes and attempts to transform the city image, much emphasis has commonly been placed on physical actions, while intangible projects have received less attention.[16]

The role of architecture in urban regeneration has often been discussed in relation to 'wow-architecture' and building new cultural infrastructure for major cultural institutions. However, one commonly used means of urban regeneration today is the use of temporary architectural intervention into city spaces, that is setting up non-permanent structures, constructions, or buildings for cultural and leisure use. In addition to urban planners and authorities, many civil associations also seek to revitalize urban spaces through event-oriented temporary architectural projects.[17] Sometimes the forms of this kind of 'pop-up architecture' can be placed in the category of urban art, community art, or 'new genre public art', a term coined by Susanne Lacy,[18] because of their artistic and communal emphases and practices involving direct social engagement. Moreover, the communal and social emphasis of recent temporary architectural projects relates them to the tradition of activist architectural practices that Nishat Awan, Tatjana Schneider, and Jeremy Till have termed 'spatial agency'.[19] These temporary event-oriented and artistic architectural projects are also closely linked to current urban culture and its aims to strengthen local communality and citizens' participation in self-made bottom-up projects. Indeed, temporary architecture has become a timely urban trend that 'can hold a subversive and anti-system potential while, paradoxically, paraphrasing the material and spatial implications of finance capitalism in today's economy', as Marina Otero Verzier notes.[20]

15 Graeme L. Evans, "Hard-Branding the Cultural City: From Prado to Prada," *International Journal of Urban and Regional Research* 27, no. 2 (2003): 417–440; Evans, "Measure for Measure," 1–25.

16 Teodoro Luque-Martinez et al., "Modeling a City's Image: the Case of Granada."

17 Kelly Chan, "Pop-up Populism: How the Temporary Architecture Craze is Changing Our Relationship to the Built Environment," *Artinfo*, May 8, 2012.

18 Susanne Lacy, "Introduction," in *Mapping the Terrain. New Genre Public Art*, ed. Suzanne Lacy (Seattle: Bay Press, 1995), 19–47.

19 Nishat Awan, Tatjana Schneider, and Jeremy Till, *Spatial Agency. Other Ways of Doing Architecture* (London: Routledge, 2011).

20 Marina Otero Verzier, "100 Urban Trends: Temporary Architecture," in *100 Urban Trends: A Glossary of Ideas from the BMW Guggenheim Lab*, 13 August 2013, accessed October 5, 2017, https://www.guggenheim.org/blogs/lablog/100-urban-trends-temporary-architecture.

The European Capital of Culture Initiative as an Instrument for Urban Regeneration

Major urban regeneration projects are long-term processes that are commonly included in the cities' development strategies aiming for long-term impacts and sustainability in economic, cultural, social, and environmental terms. Thus, urban regeneration projects involve diverse stakeholders in planning, decision-making, and implementation processes. Urban regeneration is an important political instrument through which local and regional decision-makers and authorities seek to govern a variety of economic and social issues in the cities.

During its history, the ECOC initiative has evolved from a short-term high-cultural festival to a year-long urban event that enables manifold economic and social developments and regeneration of the city space. The ethos of a high-cultural festival changed particularly after Glasgow, Scotland, was selected as the ECOC for the year 1990. Since then various ECOCs have followed Glasgow's example and used the designation as a tool for reviving declined former industrial areas by investing in different branches of culture and promoting local cultural and creative industries.[21] In addition, Glasgow's ECOC year has been described as a turning point in the alteration of the designation towards a city-marketing event.[22] Over the years the ECOC designation has become a sought-after brand that cities use for image-making, place promotion, and city marketing. Greg Richards and Robert Palmer note how in the 2010s the ECOC initiative has been particularly characterized by investments in infrastructure.[23] Following these views, Monica Sassatelli has described the current state of the initiative as a phase of 'capitalization', that is to say, cities compete to become recognized as capitals (in accordance with the name of the ECOC title), but capitalization also evokes economic capital as both a means and

21 Robert Palmer, *European Cities and Capitals of Culture. Part 1* (Brussels: European Commission & Palmer/Rae Associates, 2004), 103; Kathrin Oerters and Jürgen Mittag, "European Capitals of Culture as Incentives for Local Transformation and Creative Economies: Tendencies—Examples—Assessments," in *Whose Culture(s)?* Proceedings of the Second Annual Conference of the University Network of European Capitals of Culture held in Liverpool 16–17 Oct 2008, ed. Wim Coudenys (Pécs: The University Network of European Capitals of Culture, 2008), 88–92.

22 Oerters and Mittag, "European Capitals of Culture as Incentives," 70.

23 Greg Richards and Robert Palmer, *Eventful Cities: Cultural Management and Urban Revitalisation* (London: Routledge, 2010).

an outcome of economic processes. The ECOCs invest their assets in hope of greater economic as well as cultural returns.[24]

The ECOC designation has notably influenced urban development and transformation of the city space in several ECOCs in the Central and Eastern European countries that joined the EU in its Eastern enlargement in 2004 and 2007. The designation has previously had a similar kind of influence on cities in the so-called old EU member states. However, the transformation plans in the Central and Eastern European ECOCs have often been more comprehensive. The ECOC designation includes strong symbolic meanings and references to the idea of Europeanness—and thus the designation has been used in the 'new' EU member states as a tool for branding the city as European, indicating the city's belonging to a European cultural sphere, and, more broadly, in remapping or rethinking the cultural and political geography of Europe.[25]

In practice, the aims to regenerate urban space, invest in assets and (cultural) infrastructure, and boost economic growth through culture and cultural industries merge in the plans and policies of the ECOCs. Various ECOCs have used the initiative as a tool for reviving the city and developing its urban space, for example by upgrading cultural institutions and their facilities; modifying and modernizing squares and parks; revitalizing less-used or declined districts by preparing and cleaning their environment and installing public art in them; constructing new buildings for cultural use; renewing streets, roads and the transportation system; and renovating old estates and heritage sites. Particularly, the empty industrial estates of declined old industries close to the inner city have been transformed for use by cultural industries and as places of cultural and leisure time consumption. In spite of detailed plans, many of the large-scale construction and renovation projects in the ECOCs have not been either implemented or finished as planned due to increased expenses, financing problems, and a lack of investors.

The ECOC designation lasts only one year. Limited budgets and the short length of the designation have led many of the ECOCs to implement and host various small-scale architectural projects that influence an urban space only

---

24 Monica Sassatelli, "Europe's Several Capitals of Culture: From Celebration to Regeneration, to Polycentric Capitalization," in *The Cultural Politics of Europe. European Capitals of Culture and European Union Since the 1980s*, ed. Kiran K. Patel (London: Routledge, 2013), 64–66.

25 Tuuli Lähdesmäki, "European Capital of Culture Designation as an Initiator of Urban Transformation in the Post-Socialist Countries," *European Planning Studies* 22, no. 3 (2014): 481–497; Tuuli, Lähdesmäki, *Identity Politics in the European Capital of Culture Initiative* (Joensuu: University of Eastern Finland, 2014).

temporarily. Many of these projects aim to achieve the 'wow-effect' of spectacular architecture, but on a smaller scale and with much smaller expenses. On the one hand, the temporality and the small scale pose challenges for planning and implementation of architectural projects. On the other hand, they create possibilities for a new kind of experimentality and innovativeness regarding the material, form, location, function, and use of these constructions. In general, the development and regeneration practices in the ECOCs seek to modernize urban spaces by transforming them according to current notions of a 'good' and 'comfortable' environment and contemporary ideals of 'livable' cities.

To maximize the attention value of the ECOC designation, several cities have sought to produce new and exceptional spatial attractions in the city centers—particularly during the tourist season—in order to attract visitors, increase the local citizens' interest in their home city and its urban space, raise media attention, and have a positive influence on the city's image. These new spatial attractions have often been set up only for a couple of months and in unexpected and surprising locations in the city center. Thus, these projects can be described as architectural interventions, following the 'pop-up' ideology currently common in Western urban culture. Like pop-up restaurants, shops, and galleries in already existing buildings, pop-up architectural projects have sought to activate local citizens to experience and use urban space in a new way, bring people into less-used or unused urban spaces, and enable different people and groups to encounter each other within them.[26]

Temporary Architectural Projects in Tallinn, Turku, Maribor, and Marseille

The cultural programs in the recently designated ECOCs include several temporary architectural projects of which I have chosen four for closer examination. These cases were selected on the basis of their different regeneration objectives and concrete tactics for influencing an urban space and its uses and users. However, these cases have much in common: in all of them, the architectural interventions brought to the urban spaces, for a limited time (from a couple of months to a year), some exceptional structures that changed the meanings of the urban spaces, transformed their former uses, created a visual dialog between the elements of the spaces, and produced the possibility to experience

26 Tuuli Lähdesmäki, "Pop-up Architecture as Urban Regeneration," in *Dialectics of Space and Place across Virtual and Corporeal Topographies*, ed. by June Jordaan, Carl Haddrell, and Christine Alegria (Oxford: Inter-Disciplinary Press, 2016), 263–274.

the urban spaces in a new way. What follows is a description of the differences and similarities in the selected temporary architectural projects and a discussion on how the projects sought to function as a means of urban regeneration in their host cities. The tactics of temporality and the goals for urban regeneration in the cases are summarized at the end of the section (table 2.1).

Tallinn, the capital of Estonia, was designated as the ECOC for 2011. The biggest cultural project in the city, created particularly for the ECOC year, was the Straw Theatre—a temporary theatre building in which the Estonian Theatre NO99 and other artists performed their plays, concerts, and works of art from May to September 2011. The building, designed by Salto Architects (Maarja Kask, Karli Luik, Ralf Lõoke, and Pelle-Sten Viiburg), consisted of uncovered straw bales that were spray-painted black and supported by a wooden frame (figures 2.2 and 2.3). The theatre was located next to the Old Town on the Skoone Bastion—a neglected and decayed site that once had a wooden navy theatre and a park during the Soviet occupation of the country. After the Soviet troops left the country and the old theatre burned down, the bastion remained a closed and abandoned site that was not taken into any real-estate or recreational use, although during the past two decades various urban development plans were drafted for the site.

The temporary Straw Theatre offered a possibility to reactivate the unused location and test its potential for cultural and leisure use. The surroundings of the Straw Theatre were transformed into an open 'living room' for the local citizens and visitors. The area in front of the theatre was covered by a lawn and a wooden terrace. It included a sandpit, a hopscotch court, a grill, a snack bar, and a space where for the visitors could read cultural magazines. The venue of the theatre was described as follows in the official program book of the ECOC year of Tallinn:

> Everybody is welcome to do their morning workout there, play with their children on the playground, read intellectually enthralling magazines, eat healthy food or just listen to the birds singing and gaze at the sea.[27]

The Straw Theatre succeeded in drawing local citizens from all age groups, including families, to spend time in the recreated area. The performances in the theatre attracted a lot of visitors: both the media and the audience thanked the program for its high quality, internationalism, and topicality. The project succeeded in turning the decayed site into an urban experiment in which a

27 Andri Maimets and Maris Hellrand, eds., *Tallinn—European Capital of Culture. Stories of the Seashore. Programme* (Tallinn: Tallinn 2011 Foundation, 2010), 96.

FIGURE 2.2
Straw Theatre, designed by Salto Architects, in Tallinn, Estonia, in May 2011
PHOTO: TUULI LÄHDESMÄKI.

FIGURE 2.3
Straw Theatre (detail), designed by Salto Architects, in Tallinn, Estonia, in May 2011
PHOTO: TUULI LÄHDESMÄKI.

FIGURE 2.4 *Site of the Straw Theatre, in Tallinn, Estonia, in May 2012*
PHOTO: TUULI LÄHDESMÄKI.

'non-place' close to the city center could be experienced in a new way and given new meanings. The spectacular look and the ecological and experimental—but at the same time profoundly traditional—material of the building raised much interest in the media and among the local citizens and tourists. The wow-effect of the architecture brought a lot of people to the site just to marvel at the building.

After five months the theatre was taken down and the site was returned to its previous state (figure 2.4). The building did not leave any permanent marks on the environment; it respected the historical layers of the site, for example, by utilizing the remaining stone bases and staircases of the earlier navy theatre. Skateboarders and local youth continued to meet at the site.

Turku, a city in southwest Finland, was designated as the ECOC for 2011 together with Tallinn. The major architectural project in Turku's cultural program was the Sauna Lab. In the project, four Finnish artists (Jan-Erik Anderson, Heidi Lunabba, Harri Markkula, and Hans-Christian Berg) were invited to design site-specific artistic saunas, which were located in different sites in the city center and nearby districts for the summer months. Saunas are a profoundly traditional and living part of Finnish culture. However, the long tradition of public saunas has been diminishing during the past decades as most people bath in their own private saunas. The aim of the Sauna Lab was to bring together traditional Finnish culture and an innovative contemporary design, blur the ideas of an intimate space and a public space, revitalize the tradition of urban public saunas, and bring new communal elements to the urban space. The saunas were in public use and could be reserved for groups from June to August.

FIGURE 2.5
Temporary sauna Hot Cube, designed by Harri Markkula, in Turku, Finland, in August 2011
PHOTO: TUULI LÄHDESMÄKI.

FIGURE 2.6
Temporary sauna Solaris, designed by Hans-Christian Berg and Mika Ihanus, in Turku, Finland, in August 2011
PHOTO: TUULI LÄHDESMÄKI.

All the temporary saunas in Turku relied on unique forms and artistic elements in their design. Anderson's *Sounding Dome Sauna* was made of fiberglass and was from the outside reminiscent of a giant yellow garlic bulb. Inside the sauna, the bathers could listen to the work of sound artist Shawn Decker. The sauna was located in a local open-air swimming arena. Lunabba's *Sauna Obscura* floated on a raft in the sea near a public beach. The sauna was based on the idea of a camera obscura: the surrounding environment of the Turku Archipelago was projected upside down on the walls of the sauna and the skins of the bathers blurring the border between the inside and the outside, turning the sauna into a kind of live movie theatre. Markkula's *Hot Cube* (figure 2.5) was located in the city center above the Aura River, which flows through the city. The sauna was made using tarred wooden panels that formed a windowless cube. The river could be seen through lattices in the sauna floor. Breg's *Solaris* (figure 2.6), designed in cooperation with industrial designer Mika Ihanus, was a transparent round glass building located on a small hill next to one of the city's art museums, a busy road, and a bridge crossing the Aura River.

The Sauna Lab brought unexpected functions to the urban space by offering the possibility of bathing and relaxing in the urban space, and meeting other citizens and cultural tourists within an intimate space. The artistic design of the saunas sparked much interest in the media and among citizens and tourists. According to the bathers, bathing in an artistically unique sauna in an unexpected place in the city was experienced as extraordinary and unforgettable. In September 2011 the saunas were removed: some of them were offered for sale and some of their prototypes were introduced for further production. All the saunas have reappeared in some other venue or have visited other cities as a part of some event or cultural festival.

Whether influenced by the *Sauna Lab* or not, during the past few years the tradition of public saunas has undergone a rapid recovery in Finland, as two public 'design' saunas have been set up in the center of the capital city Helsinki. Their popularity among the local people and tourists has recently inspired launches of similar plans in other major Finnish cities, as well.

Maribor, a city in Northern Slovenia, was designated as one of the ECOCs for 2012. During the year one of the main squares in the city, the Trg Svobode, functioned as the location of a temporary exhibition space called White Noise (figure 2.7). It was designed by Austrian architecture collective SOMA (Johan Tali, Alex Matl, and Karin Dobbler), which in 2011 won an open architectural competition held by the Salzburg Regional Government with the purpose of promoting contemporary architecture and architectural interventions into public space. The pavilion was constructed of 1500 aluminum bars, which were connected following a set of rules and variations to form an organic sculpture-like structure. The aluminum structure of the pavilion was lined with a tarp.

After the win, the SOMA architects toured various cities with the pavilion. Maribor bought it for six weeks as a part of the city's ECOC program. From September to October 2012 the pavilion hosted diverse small concerts, exhibitions, discussions, and screenings in the city (figure 2.8). The look of the pavilion was extraordinary and eye-catching, forming a strong contrast with the historical buildings surrounding the square. It also brought new cultural uses to the square, which is used for fairs and as a market place and on which is located a large bronze monument of the National Liberation War.

The White Noise pavilion is an example of an international, temporary, and movable artistic and architectural project that can be situated for a limited period of time in an urban space to bring it to life in a new way and to activate cultural and communal participation by the citizens. Similar kinds of 'touring'

FIGURE 2.7
White Noise, designed by the SOMA collective, in Maribor, Slovenia, in October 2012
PHOTO: TUULI LÄHDESMÄKI.

FIGURE 2.8
Exhibition 'Ten' by Nenad Cizl in White Noise, designed by the SOMA collective, in Maribor, Slovenia, in October 2012
PHOTO: TUULI LÄHDESMÄKI.

projects have been utilized by several recent ECOCs as part of their cultural programs.[28]

Marseille, the French city on the shore of the Mediterranean Sea, was designated as one of the ECOCs for 2013. The city authorities commissioned vast development plans for the cultural year and implemented various costly regeneration projects in which public spaces in the city center were renovated and new impressive buildings were constructed. Especially the Museum of European and Mediterranean Civilisations (MuCEM) designed by architect Rudy Ricciotti and the conference center Villa Méditerranée designed by architect Stefano Boeri can be regarded as 'wow-architecture'.

In addition to these permanent architectural projects, one of the core spots in Marseille during its ECOC year was a temporary information center and cultural venue M Pavilion (figure 2.9) designed by Philippe Malta. The pavilion was built over an extension of the city hall and it thereby utilized an already-existing architectural space. The extension of the city hall of Marseille was commissioned by the city government from architect Franck Hammoutène, who won the competition on the project in 1999. The extension was completed in 2006, following the recently much-discussed spatial ideas of underground architecture:[29] new municipal offices, meeting rooms, exhibitions

28 For example, the Flying Grass Carpet, a project by Dutch HUNK-Design and ID-Eddy, which consists of a combination of several artificial grasses with different characteristics and colors woven together to resemble a huge Persian carpet, visited in 2010 three ECOCs: Pécs (Hungary), Essen (Germany), and Istanbul (Turkey). The Colourscape, a project by English Eye Music Trust comprising a huge walk-in network of interconnected chambers that bend the colors of light, visited Turku in August 2011.

29 Ernst von Meijenfeldt et al., *Below Ground Level: Creating New Spaces for Contemporary Architecture* (Basel: Birkhäuser, 2003).

spaces, and a large foyer covering altogether 8300 square meters were located under the ground level next to the old 17th century city hall. Above the extension, a new public space was created by covering and refurbishing the Villeneuve-Bargemon square and by letting the square follow the original form of the landscape, sloping towards the nearby Old Port.

The M Pavilion was constructed partly in front of the entrance to the extension of the city hall and partly above it by utilizing the its foyer, exhibition spaces, and large meeting room as part of the pavilion. The temporary construction of the pavilion covered 1200 square meters, forming a wave-like shape on the side of the square. The pavilion was built using a wooden framework supporting large sheets of polycarbonate, which let daylight into the building to create a transparent impression (figure 2.10). The fundamental idea in its construction was sustainability: when dismantled, all of its components could be reused or recycled.

FIGURE 2.9
The M Pavilion, designed by Philippe Malta, in Marseille, France, in October 2013
PHOTO: TUULI LÄHDESMÄKI.

FIGURE 2.10
Lobby of the M Pavilion, designed by Philippe Malta, in Marseille, France, in October 2013
PHOTO: TUULI LÄHDESMÄKI.

The M Pavilion was opened in January 2013 and dismantled in February 2014. During that time, 1.2 million people visited it.[30] Its main purpose was to function as a showcase venue for the cultural program of Marseille's ECOC year

TABLE 2.1 *Tactics of temporality in the architectural case projects and their goals for urban regeneration*

Cases	Tactics of temporality	Goals for urban regeneration
The Straw Theatre (Tallinn)	· ecological, traditional, non-permanent construction material · utilization of previous structures on the site as a part of the building	· to enliven a less-used and decayed site near the city center with an attractive design · to create an 'urban living room' for the citizens · to provide a new venue for cultural performances
Sauna Lab (Turku)	· small-scale architectural projects of which some were sold and some introduced for further production	· to create new surprising use for urban spaces · to merge architecture and art · to bring new communal activities to the city center
White Noise (Maribor)	· a movable architectural project that can be bought to a site for a short duration	· to provide a new attractive venue for cultural performances
The M Pavilion (Marseille)	· constructed from materials and with techniques that enabled dismantling of the building and recycling of its elements · utilization of current architectural space on the site	· to create a new meeting place for the citizens in the city center · to provide a new venue for cultural performances

30 "MP13: le Pavillon M a accueilli 1,2 million de visiteurs," *La Provence.com*, 31 December 2013, accessed May 30, 2014, http://www.laprovence.com/actu/leco-en-direct/2688098/mp13-le-pavillon-m-a-accueilli-12-million-de-visiteurs.html.

and communicate with the citizens and visitors about the city. The pavilion succeeded in serving as a meeting place in the city, spurring several citizens to wish at the end of the ECOC year for the legacy of the pavilion to be continued and a new building dedicated to culture to be set up in the city.

Contextualizing Recent Temporary Architecture

The current interest in regenerating urban spaces by setting up temporary buildings, such as the case projects of this study, reflects recent theoretical discussions on and conceptualizations of urban planning. The focus of these discussions is on the idea and ideal of increasing interaction and communication between citizens in these urban spaces, strengthening communality and urban identities in the city, rediscovering the city center and its activities, and fostering human-scale and pedestrian-friendly urban design as a basis for 'livable' cities.

Nigel Taylor has identified a paradigmatic change in urban planning theories and practices during the 1990s. According to him, at that time views in urban planning started to follow a new communicative paradigm. It meant a new emphasis on interaction and communication instead of 'pure' rational, scientific, and technical thinking as the bases for managing urban planning.[31] The emphasis on interaction and communication easily incorporated cultural points of view into the aims and ideas of urban planning. Indeed, cultural planning has become one of the key points in urban planning discourses. These kinds of ideas and emphases in urban planning have also been described as characterizing the shift from modern to post-modern urban planning principles.[32] The ideals of post-modern urban planning are considered to include, for example, fostering of urban identities and appreciation of history and traditions of places, participatory planning methods, and pursuit of human-scale and compact urban forms.[33] Moreover, similar ideas characterize the planning principles conceptualized as New Urbanism. It stresses, for example, a rediscovery of the city center and its activities, pedestrian-friendly urban design, diversity and openness of the public space, urban esthetics, quality of design, and communality and interaction of citizens in order to increasingly enliven the urban environment.[34]

31 Nigel Taylor, *Urban Planning Theory Since 1945* (London: Sage, 1998).

32 Sonia A. Hirt, "Towards Postmodern Urbanism? Evolution of Planning in Cleveland, Ohio," *Journal of Planning Education and Research* 25, no. 1 (2005), 29.

33 Hirt, "Towards Postmodern Urbanism?"

34 Tigran Haas, ed., *New Urbanism & Beyond. Designing Cities for the Future* (New York: Rizzoli, 2008).

The recent ideas of urban planning intertwine with the economic and social motives described in the second section of this chapter. Besides cultural and creative economy, the current aims of urban planning reflect the ideas brought together in the concept of experience economy. Some scholars have described these ideas as 'economy of fascination' in order to emphasize how cities have been competitively themed to market their unique 'experience world'.[35] During the past three decades, politicians, planners, and marketing experts have increasingly focused on developing experiences in order to foster consumption.[36] As a consequence, cities have started to seek possibilities to offer people spectacular urban environments, designed for these moments of extraordinary experiences, to profit from their interest in gaining new experiences.[37] Experience-focused urban planning has moved more and more from investing in 'hard' location factors, such as major buildings, towards 'soft' location factors, such as recreational activities and place-based installations.[38] Estheticization of an urban space is at the same time a means and an outcome of experienced-focused urban planning. On the one hand, estheticization is a process that seeks to create novelty, surprise, and excitement, but on the other hand, it may only result in superficial changes.[39] As Anne-Marie d'Hauteserre states:

> The aestheticisation of urban neighbourhoods represents a concern with the production of difference relative to the visibility of phenomena, mustering emotion and affect through iconic images and spectacle, creating a dreamscape for visual consumption, a spectacular experience.[40]

Many of the ECOCs have sought to boost tourism in the cities by offering visitors diverse recreational and cultural activities and spectacular urban places. Daniel Habit claims that the urban spaces in the ECOCs have even been turned

35 For example, Heiko Schmid, *Economy of Fascination. Dubai and Las Vegas as Themed Urban Landscapes* (Berlin: Gebrüder Borntraeger, 2009).

36 Doreen Jakob, "The Eventification of Place: Urban Development and Experience Consumption in Berlin and New York City," *European Urban and Regional Studies* 20, no. 4 (2013), 447–459.

37 Anne-Marie d'Hauteserre, "Val d'Europe: A Pioneering Turn to 'Experience' Planning?" *European Urban and Regional Studies* 20, no. 4 (2013), 435–446.

38 Peter Allingham, "Art, Media, and Sense-making in Responsive Urban Environments," *Rask* 40 (2014), 31–52.

39 d'Hauteserre, "Val d'Europe: A Pioneering Turn to 'Experience' Planning?" 437.

40 Ibid.

into 'citytainment'.[41] He even compares the implementation of the ECOC year of Sibiu in Romania in 2007 to the process described as disneyfication.[42] David Harvey has used the concept of disneyfication to describe attempts to transform the urban environment into safe and clean spaces of entertainment, which promote the city and attract capital, tourists, and people who share similar values.[43] The logic of Disneyland has globally influenced various initiatives and attempts to organize leisure activities and build environments for them.[44]

Conclusions

The explored cases from the four ECOCs indicate that temporary architectural projects can revitalize, regenerate, and produce new meanings for urban spaces, activate people to use public spaces in a new way, increase social interaction in the city, and raise positive public attention. Temporary architectural projects are easy, fast, cheap, and less risky modes of creating wow-effects within urban spaces—an outcome that urban planners often long for—than large-scale 'wow-architecture'. On the one hand, the cultural, social, and economic impacts of temporary projects may be short-term and limited in scale. On the other hand, temporality respects the existing structures and historical layers of the city without seeking to change the urban space permanently. Temporality enables implementation of innovative architectural and communal experiments, playing with the experiential and affective dimensions of the urban space, and fast and flexible reaction to changing social dynamics in the city.

Contemporary views in urban regeneration particularly emphasize the role of local communities in planning and implementing regeneration projects. As a consequence, social and community regeneration has been recognized as an integral part of urban regeneration.[45] Today, the success of urban regeneration projects is not only measured by physical transformation of the city space or its economic impact, but also by its outcomes in creating vibrant communities,

41 Daniel Habit, "Pheripheral ECOCs between Cultural Policy and Cultural Governance: The Case of Sibiu 2007," in *The Cultural Politics of Europe. European Capitals of Culture and European Union Since the 1980s*, ed. Kiran K. Patel (London: Routledge, 2013), 132.

42 Id., 134.

43 David Harvey, *The Condition of Postmodernity* (Oxford: Basil Blackwell, 1989).

44 John Hannigan, *Fantasy City. Pleasure and Profit in the Postmodern Metropolis* (London: Routledge, 1998).

45 Rachel Granger, "Social and Community Issues," in *Urban Regeneration*, eds. Peter Roberts, Hugh Sykes, and Rachel Granger (London: Sage, 1999), 107–110.

interaction between citizens, a feeling of comfort in the urban space, active participation in cultural and social issues, and interest in the urban environment. Attempts to enliven city centers and make cities more 'livable' often originate from top-down decision-making and urban planning practices—of which the ECOC initiative functions as an example. Scholars have criticized these practices for forgetting to involve local people and ordinary citizens in the urban planning and regeneration processes.[46]

Temporary small-scale experimental architectural projects could offer possibilities for more effective inclusion of local people in communal and cultural development and regeneration of the urban environment. Pop-up ideology enables the citizens to plan and implement grass-roots and bottom-up initiatives and create temporary venues for communal and cultural activities. The future of the ECOC initiative will show whether the possibilities of temporary architecture will be better utilized in urban and community regeneration and whether the bottom-up initiatives planned and implemented by the citizens themselves will include temporary space-making and architectural projects that enliven the city.

Bibliography

Allingham, Peter. "Art, Media, and Sense-making in Responsive Urban Environments." *Rask* 40 (2014): 31–52.

Awan, Nishat, Tatjana Schneider, and Jeremy Till. *Spatial Agency. Other Ways of Doing Architecture*. London and New York: Routledge, 2011.

Beauregard, Robert and Briavel Holcomb. *Revitalizing Cities*. Washington, DC: Association of American Geographers, 1981.

Bonet, Lluís and Emmanuel Négrier. "The Ends(s) of National Cultures? Cultural Policy in the Face of Diversity." *International Journal of Cultural Policy* 17, no. 5 (2011): 574–589.

Chan, Kelly. "Pop-up Populism: How the Temporary Architecture Craze is Changing Our Relationship to the Built Environment." *Artinfo*, May 8, 2012. http://www.artinfo.com.

Cowan, Robert. *The Dictionary of Urbanism*. Tisbury: Streetwise Press, 2005.

46 Tim Hall, "Public Art, Civic Identity and the New Birmingham," in *Remaking Birmingham: The Visual Culture of Urban Regeneration*, ed. Liam Kennedy (London: Routledge, 2004), 71; Beatriz García, "Cultural Policy and Urban Regeneration in Western European Cities: Lesson from Experience, Prospects for the Future," *Local Economy* 19, no. 4 (2004): 312–326; Evans, "Measure for Measure," 1–25.

Evans, Graeme L. "Hard-Branding the Cultural City: From Prado to Prada." *International Journal of Urban and Regional Research* 27, no. 2 (2003): 417–440.

Evans, Graeme L. "Measure for Measure: Evaluating the Evidence of Culture's Contribution to Regeneration." *Urban Studies* 42, no. 5–6 (2005): 1–25.

García, Beatriz. "Cultural Policy and Urban Regeneration in Western European Cities: Lesson from Experience, Prospects for the Future." *Local Economy* 19, no. 4 (2004): 312–326.

Granger, Rachel. "Social and Community Issues," In *Urban Renegeration*, edited by Peter Roberts, Hugh Sykes, and Rachel Granger, 99–112. London: Sage, 1999.

Haas, Tigran, ed. *New Urbanism & Beyond. Designing Cities for the Future*. New York: Rizzoli, 2008.

Habit, Daniel. "Pheripheral ecocs between Cultural Policy and Cultural Governance: The Case of Sibiu 2007." In *The Cultural Politics of Europe. European Capitals of Culture and European Union Since the 1980s*, edited by Kiran K. Patel, 127–140. London: Routledge, 2013.

Hall, Tim. "Public Art, Civic Identity and the New Birmingham." In *Remaking Birmingham: The Visual Culture of Urban Regeneration*, edited by Liam Kennedy, 63–71. London: Routledge, 2004.

Hannigan, John. *Fantasy City. Pleasure and Profit in the Postmodern Metropolis*. London: Routledge, 1998.

Harvey, David. *The Condition of Postmodernity*. Oxford: Basil Blackwell, 1989.

d'Hauteserre, Anne-Marie. "Val d'Europe: A Pioneering Turn to 'Experience' Planning?" *European Urban and Regional Studies* 20, no. 4 (2013): 435–446.

Hirt, Sonia A. "Towards Postmodern Urbanism? Evolution of Planning in Cleveland, Ohio." *Journal of Planning Education and Research* 25, no. 1 (2005): 27–42.

Horstmannshoff, Kai. *The Loop. Chicago Architecture and the Social Imaginary*. Bielefeld: Transcript, 2018.

Jakob, Doreen. "The Eventification of Place: Urban Development and Experience Consumption in Berlin and New York City." *European Urban and Regional Studies* 20, no. 4 (2013), 447–459.

Lacy, Susanne. "Introduction." In *Mapping the Terrain. New Genre Public Art*, edited by Suzanne Lacy, 19–47. Seattle: Bay Press, 1995.

Lähdesmäki, Tuuli. "European Capital of Culture Designation as an Initiator of Urban Transformation in the Post-Socialist Countries." *European Planning Studies* 22, no. 3 (2014): 481–497.

Lähdesmäki, Tuuli. *Identity Politics in the European Capital of Culture Initiative*. Joensuu: University of Eastern Finland, 2014.

Lähdesmäki, Tuuli. "Pop-up Architecture as Urban Regeneration." In *Dialectics of Space and Place across Virtual and Corporeal Topographies*, edited by June Jordaan, Carl Haddrell, and Christine Alegria, 263–274. Oxford: Inter-Disciplinary Press, 2016.

La Provence.com. "MP13: le Pavillon M a accueilli 1,2 million de visiteurs." December 31, 2013. Accessed May 30, 2014. http://www.laprovence.com/actu/leco-en-direct/2688098/mp13-le-pavillon-m-a-accueilli-12-million-de-visiteurs.html.

Luque-Martinez, Teodoro, Salvador Del Barrio-García, José Ángel Ibáñes-Zapata, and Migual Ángel Rodriguez-Molina. "Modeling a City's Image: the Case of Granada." *Cities* 24, no. 5 (2007): 335–352.

MacCannell, Dean. *The Tourist. A New Theory of the Leisure Class*. New York: Schocken Books, 1989.

Maimets, Andri and Maris Hellrand, eds. *Tallinn—European Capital of Culture. Stories of the Seashore. Programme*. Tallinn: Tallinn 2011 Foundation, 2010.

von Meijenfeldt, Ernst, et al. *Below Ground Level: Creating New Spaces for Contemporary Architecture*. Basel: Birkhäuser, 2003.

Näss, Hans Erik. "The Ambiguities of Intercultural Dialogue: Critical Perspectives on the European Union's New Agenda for Culture." *Journal of Intercultural Communication* 23 (2010). Accessed October 5, 2017. https://www.immi.se/intercultural/nr23/nass.htm.

Oerters, Kathrin and Jürgen Mittag. "European Capitals of Culture as Incentives for Local Transformation and Creative Economies: Tendencies—Examples—Assessments." In *Whose Culture(s)?* Proceedings of the Second Annual Conference of the University Network of European Capitals of Culture held in Liverpool 16–17 Oct 2008, edited by Wim Coudenys, 70–97. Pécs: The University Network of European Capitals of Culture, 2008.

Otero Verzier, Marina. "100 Urban Trends: Temporary Architecture." In *100 Urban Trends: A Glossary of Ideas from the BMW Guggenheim Lab*. August 13, 2013. Accessed October 5, 2017. https://www.guggenheim.org/blogs/lablog/100-urban-trends-temporary-architecture.

Paddison, Ronan. "City Marketing, Image Reconstruction and Urban Regeneration." *Urban Studies* 30, no. 2 (1993): 339–350.

Palmer, Robert. *European Cities and Capitals of Culture. Part I*. Brussels: European Commission & Palmer/Rae Associates, 2004.

Richards, Greg and Robert Palmer. *Eventful Cities: Cultural Management and Urban Revitalisation*. London: Routledge, 2010.

Sassatelli, Monica. "Europe's Several Capitals of Culture: From Celebration to Regeneration, to Polycentric Capitalization." In *The Cultural Politics of Europe. European Capitals of Culture and European Union Since the 1980s*, edited by Kiran K. Patel, 55–71. London: Routledge, 2013.

Schmid, Heiko. *Economy of Fascination. Dubai and Las Vegas as Themed Urban Landscapes*. Berlin: Gebrüder Borntraeger, 2009.

Taipale, Kaarin. *Cities for Sale. How Economic Globalization Transforms the Local Public Sphere*. Centre for Urban and Regional Studies Publications. Studies in Architecture A 36. Espoo: Helsinki University of Technology, 2009.

Taylor, Nigel. *Urban Planning Theory Since 1945*. London: Sage, 1998.

Temelová, Jana. "Urban Revitalization in Central and Inner Parts of (Post-socialist) Cities: Conditions and Consequences." In *Regenerating Urban Core*, edited by Tuomas Ilmavirta. Publications in the Center for Urban and Regional Studies C72, 12–25. Espoo: Helsinki University of Technology, 2009.

Vitruvius. *The Ten Books on Architecture*. Translated by Morris Hicky Morgan. Cambridge: Harvard University Press, 1914.

Zielenbach, Sean. *The Art of Revitalization. Improving Condition of Distressed Inner-City Neighborhoods*. London: Garland Publishing, 2000.

CHAPTER 3

After the Event: Considering Media Images and Skyscraper Architecture at the Turn of the 21st Century

Sanja Rodeš

Introduction

This chapter explores the relationship between the selected examples of skyscraper architecture and media images in the aftermath of the 9/11 terrorist attack. 9/11 is positioned as an event which is architecturally represented by, and inextricably related to, the architecture of the Twin Towers, further granting that architecture different political, social, and economic agencies.[1] The focus of the chapter is on the role of the media and image in facilitating these multiple agencies for architecture, and it further reflects on the production of two other skyscraper projects and their assumed relationship with 9/11's media image. These projects are the new World Trade Center (the new WTC, hereafter), initially designed by Daniel Libeskind in New York (2006–present) in the US, and the new China Central Television Headquarters (the CCTV, hereafter) in Beijing in China, designed by Office for Metropolitan Architecture (OMA, hereafter), which opened in 2012. Considering that the CCTV and the new WTC have a relationship with image—both in their conception and as final projects—this chapter asserts that image figures as crucial for understanding their architecture, and should be examined as such. These projects also relate to the image of the 9/11 media event (as is demonstrated in this chapter), in the sense that this event acted as a catalyst for media attention. The aim of the chapter is to examine productively the relationships between the selected architectural examples in the public sphere and media image, arguing that this relationship with the public and image is crucial to understanding their architecture. It analyses these relationships against the concepts of 'iconomy', as

1 Keller Easterling, "Enduring Innocence," *Grey Room* 7, Spring, Special issue: On 9/11 (2002), 107; Andrew Herscher, "The Language of Damage," *Grey Room* 7, Spring, Special issue: On 9/11 (2002), 70; Reinhold Martin, "One or More," *Grey Room* 7, Spring, Special issue: On 9/11 (2002), 115–117.

 | DOI 10.1163/9789004376793_004

proposed by art historian Terry Smith (2006), and 'translation', as proposed by Edward W. Said (1983), who was a professor of literature at Columbia University.[2] The chapter argues that these concepts evidence the circulation of (architectural) images in the public sphere and as such are able to reveal more about the relationships between selected architecture and media images.

Often quoted as examples of iconic architecture in architectural literature, skyscrapers are frequently imbued with a symbolism that is beyond simply architectural, and that maintains a relationship with image and media—these buildings become indicators of political structures or economic power. Considering this symbolism, this chapter analyses how 9/11 has positioned architecture within globalized public imaginaries and 'iconomy'.[3] This global imaginary refers to the changes globalization brought to the 'social imaginary', explained by Manfred B. Steger, as the mode that 'provides the most general parameters within which people imagine their communal existence.'[4] Steger continues that such social imaginary was, post Second World War, influenced by the influx of the ideas, images, and people brought into the national imaginary by new technologies. Such ideas, images, and people constituted the 'global', influencing and undermining national identities.[5]

2 Terry Smith, *The Architecture of Aftermath* (Chicago: University of Chicago Press, 2006), 1–12; Edward W. Said, "Traveling Theory," in Edward W. Said, *The World, the Text, and the Critic* (Cambridge: Harvard University Press, 1983), 226–247.

3 For analysis of the Twin Towers destruction within iconomy, see Smith, *The Architecture of Aftermath*, 1–12.

4 Steger is a professor and former director of Globalism Research Centre at Royal Melbourne Institute of Technology (RMIT) in Australia. Introduction to Manfred B. Steger, *The Rise of the Global Imaginary: Political Ideologies from the French Revolution to the Global War on Terror* (Oxford; New York: Oxford University Press, 2008), 6. Steger quotes the philosopher Charles Taylor, emphasizing 'that the social imaginary is neither a theology nor an ideology, but an implicit "background" that makes possible communal practices and a widely shared sense of their legitimacy. It offers explanation how "we"—the members of the community—fit together, how things go on between us, the expectations we have of each other, and the deeper normative notions and images that underlie those expectations'. See Charles Taylor, *Modern Social Imaginaries*, Series Public Planet Books (Durham: Duke University Press, 2004), 2, 23–26; Charles Taylor, *A Secular Age* (Cambridge: Belknap Press of Harvard University Press, 2007), 423–538, quoted in Steger, *The Rise of the Global Imaginary*, 6.

5 In the 1990s, the globalization process was referred to on mostly economic terms, related to 'the creation of a single global free market and spread of consumerist values around the world [...] they [global social elites] translated the rising social imaginary into largely economistic claims laced with references to the global'. See Steger, *The Rise of the Global Imaginary*, 11.

In addition, the relevant concept for the analysis in this chapter is iconomy, which is initially proposed in *The Architecture of Aftermath* (2006) by Smith.[6] He sees iconomy as complex phenomena bound to 'image economy',[7] related to the process in which images attract attention and are further exchanged for economic value. More importantly, iconomy includes 'exchanges between people, things, ideas, interest groups, and cultures that take predominantly visual form.'[8] Smith acknowledges the powerful exchange network facilitated by image flows: 'the competition between visual images … has spread to saturate most of the ways in which peoples, nations, religions, even civilizations relate to each other, not simply at the level of stereotype but at the deepest levels of psyche and society.'[9] These image flows contribute to the creation of visual stereotypes and define how different cultures relate to each other. Smith argues for the significance of such image flows by noting the cultural exchange able to define identity, build shared values, and be 'creatively enriching'.[10] Iconomy is relevant for this chapter primarily because it acknowledges the importance of images in contemporary architecture; it also positions architecture as part of broader social, political, or economic contexts, and as a part of the public sphere. This chapter adopts the concept of iconomy in analyzing the (architectural) image flows and their wider impact on public (as identified by Smith), and focuses on examining iconomy of selected architectural projects in the public sphere and their relationship with media images.

The second relevant concept—translation—relies on the existence of iconomy, and is also expected to reveal more about these image flows. Translation is primarily relevant for this chapter for clarifying the influence of architectural media image on the creation of architecture, or on its production. In its most obvious terms, translation denotes the linguistic crossing of a text from one language into another. In the words of philosopher and cultural critic Walter Benjamin, 'translation thus ultimately serves the purpose of expressing the central reciprocal relationship between languages.'[11] While acknowledging that translation commonly represents the relationship between two languages in communicating the same content, translation necessarily appears in other,

6 Smith, *The Architecture of Aftermath*, 1–2.

7 Ibid.

8 Id., 2.

9 Id., 5.

10 Ibid.

11 Walter Benjamin, "The Task of the Translator: An Introduction to the Translation of Baudelaire's *Tableaux Parisiens*", in Walter Benjamin and Hannah Arendt, *Illuminations*, trans. Harry Zohn (London: Fontana Press, 1992), 73.

more abstract forms. In this chapter, it denotes the translation of architecture into image, or translation of the idea into a different one, or translation of one image into another. All architectural examples proposed in this chapter are defined with the translation of their architecture into a media image of high return, which Smith calls 'iconotype'.[12] Admitting that the translation of architecture into image is a part of the process of translating ideas, this chapter focuses on translating ideas and concepts as particularly significant. Moreover, translating one architectural image into another necessarily reflects (and is the consequence of) the translation of ideas as well. In this manner, the chapter considers the circulation of architectural images within different processes of translation, aiming to also explore the role architectural media image has in processes related to the translation of architectural concepts and ideas.

Said identifies stages of the transfer of ideas, proposing stages of translation related to literary ideas. The first stage is a point of origin, where the idea is created. The second stage is the distance travelled. The third is a group of conditions that confront the transplanted idea. Finally, the fourth stage is the fully or partly accommodated idea somewhat transformed by a new context and users.[13] The relevance of this analysis is that it equally relates to architectural ideas. Moreover, as Christoph Schnoor reminds us, it is relevant to note that translation 'is not simply a mechanical act of transferring an idea into a new realm but a creative act.'[14] This creative act is of interest to this chapter, in which the translation of a particular architectural idea, image, or architecture via the vehicle of media is significant for producing the architectural projects examined in this chapter. This chapter proposes that both iconomy and translation are media-facilitated processes, and as such relevant for the analysis of these projects and their relationship with media image.

The chapter first considers 9/11 as a media event, arguing for its crucial and intentional relationship with the media. It further analyses the iconomy of the Twin Towers as it relates to that event, and the translation of the Twin Towers media image. It examines two types of translation: the first involves translation of post-9/11 extra-architectural agencies into the new WTC architecture. This translation is linked to the notion of 'limit', as proposed by architect Bernard

12 Smith, *The Architecture of Aftermath*, 22.

13 Said, "Traveling Theory," 226–227.

14 Christoph Schnoor is an associate professor in architecture at University of Technology in Auckland, New Zealand. Christoph Schnoor, "Foreword," in *Proceedings of the Society of Architectural Historians, Australia and New Zealand: 31, Translation*, ed. Christoph Schnoor (Auckland, New Zealand SAHANZ and Unitec ePress; and Gold Coast, Queensland: SAHANZ, 2014), 5.

Tschumi, where the 9/11 media event is proposed as the 'limit' of the new WTC architecture.[15] This chapter considers both the destroyed and new WTC within their post-9/11 media contexts, arguing that 9/11 became inseparable from understanding their architecture. More importantly, the chapter asserts that 9/11 has positioned the new WTC, in particular, within the public sphere, and that this relationship with the public is crucial to understanding its architecture. For example, after 9/11, the new WTC project immediately became the focus of media and public attention. The destruction of the Twin Towers changed the socio-political context of the WTC site, inducing an overall atmosphere of anxiety and mourning, along with an immediate need to assign the site new meaning.[16]

The second translation examined in this chapter is identified in the relationship between the 9/11 iconotype and the concept of the CCTV. This is evident in OMA's publication *Content: Triumph of Realization* (2004), where architect Rem Koolhaas correlates and compares the CCTV with the Twin Towers.[17] While arguing that the only genius of these towers was their twinning, Koolhaas based the CCTV on the concept of the Twin Towers in order to experiment with the contemporary skyscraper form.[18] The development of the CCTV form and concept is presented in *Content* under the heading 'Universal Modernization Patent—Skyscraper Loop (2002)'. Here, the authors demonstrate the conceptual progress from Twin Towers to skyscraper loop, calling it 'Patent for "Bent Skyscraper"'.[19] The chapter argues that one architectural image (the destruction of the Twin Towers) was directly translated into another (the CCTV). This also represents the translation of an idea, where translating the architectural concept of the Twin Towers relates to the discourse of reinventing the skyscraper, as Koolhaas proposes in *Content*.[20] This chapter thereby examines the CCTV in both its physical and media contexts to learn more about the relationships this project has with image and media, and by extension, it expands the understanding of the relations between media images of 9/11 and both WTC and CCTV.

---

15 Tschumi explains such a limit as a synonym for the 'boundaries' that define architecture; see Bernard Tschumi, *Architecture and Disjunction* (Cambridge: MIT Press, 1994), 101–102.

16 Marita Sturken, "The Aesthetics of Absence: Rebuilding Ground Zero," *American Ethnologist* 31, no. 3 (2004).

17 Rem Koolhaas is one of the leading architects in OMA. See Rem Koolhaas and Brendan McGetrick, *Content: Triumph of Realization* (Köln: Taschen, 2004).

18 Koolhaas and McGetrick, *Content*, 237.

19 See Koolhaas and McGetrick, *Content*, 511.

20 Ibid.

9/11: Planned as a Media Event

After 9/11, discourses appeared in architectural literature, media theory, and philosophy regarding the relationship between this event with the media, arguing that an extensive media presence was one of the guiding goals of the event.[21] Media theory examined the event as an example of a 'disaster marathon', a term that refers to the event's duration and its continuous presence on television and radio, with scenes replayed in order to attract the attention of the widest possible audience.[22] The visual effect of 9/11, also referred somewhat disturbingly as 'spectacular', relied on the continuous mediatization of the city's tallest towers collapsing, which, as philosopher Slavoj Žižek explains it, could have been mistaken for a Hollywood blockbuster movie.[23] It attracted extensive attention for its element of surprise, and for being a disruptive media event.[24] The direct relationship between this event and the media was also due to pre-planning on the part of the terrorists. The selected city for this terrorist attack was one of the most iconic in the media, and the event's timing allowed for global audiences in all continents to be 'present' in front of the television as it unfolded.[25] Therefore, 9/11's spectacular effect and extensive mediatization is not surprising, considering that it was pre-planned by its perpetrators.

In architectural literature, Smith, in his essay 'Media as Missile', also notes that 9/11 was consciously planned to be a media event. It 'involved at least four planes, and at least four targets, all to be hit roughly simultaneously, or in close sequence, and in ways that would command maximal attention, that would

21 Smith, *The Architecture of Aftermath*, 140–142; Slavoj Žižek, *Welcome to the Desert of the Real* (New York, NY: Wooster Press, 2001), 11–12; Elihu Katz and Tamar Liebes, "'No More Peace!': How Disaster, Terror and War Have Upstaged Media Events," *International Journal of Communication* 1 (2007), 161; Menahem Blondheim and Tamar Liebes, "Live Television's Disaster's Marathon of September 11 and Its Subversive Potential," *Prometheus* 20, no. 3 (2002); Agnizeszka Stepinska, "9/11 and the Transformation of Globalized Media Events," in *Media Events in a Global* Age, eds. Nick Couldry, Andreas Hepp and Friedrich Krotz (London; New York: Routledge, 2010), 210.

22 Tamar Liebes, James Curran, and Elihu Katz (eds.), *Media, Ritual, and Identity*, Series Communication and Society (London; New York: Routledge, 1998), 7.

23 Žižek, *Desert of the Real*, 15–17.

24 For further information on the media event categories, see Daniel Dayan and Elihu Katz, *Media Events: The Live Broadcasting of History* (Cambridge: Harvard University Press, 1992), 1–10, 68–73, and Couldry, Hepp, and Krotz, *Media Events in a Global Age*, 8.

25 Katz and Liebes, "'No More Peace!'", 161; for further information, see Blondheim and Liebes, "Live Television's Disaster's Marathon."

create the most spectacular effect within a spectacle saturated world.'[26] He quotes Žižek, one of many to note the same relationship with the media regarding 9/11: 'the "terrorists" themselves did not do it primarily to provoke real material damage, but for the spectacular effect of it.'[27] While seven buildings (1 and 2 WTC, which were the Twin Towers, along with WTC 3, 4, 5, 6, and 7) were destroyed in 9/11 or demolished afterwards, the Twin Towers constituted the iconotype, since they were the tallest, and therefore produced a media spectacle once they collapsed.[28]

The relationship between the Twin Towers and the media, as related to visual culture, is also noted by visual culture theorist Nicholas Mirzoeff, who states that we live 'in the world understood by such metonyms as 9/11, Abu Ghraib and Hurricane Katrina.'[29] Mirzoeff suggests that 9/11 was unlikely to produce a favorable impact for those who supported the perpetrators. He emphasizes that their intention was 'pure visibility',[30] and, as one can also argue—image. The power of such a media image is visible in the chain of political violence that ensued in the event's aftermath. For example, the bombings in Madrid in 2004 and in London in 2005—one of the terrorists in the London attack allegedly admitted that he was inspired by seeing 9/11 on television.[31] Mirzoeff therefore acknowledges the influence of 9/11's media image on these successive terrorist attacks, where 9/11, primarily an act of terrorism, inspired further terrorism. The power of image to influence further production of both terrorism and architecture is relevant for its impact on public consciousness, which

---

26 Smith, *The Architecture of Aftermath*, 140.

27 Žižek, *Desert of the Real*, 11; Smith, *The Architecture of Aftermath*, 141; see also Reporters, Writers, and Editors of *Der Spiegel* Magazine, *Inside 9–11: What Really Happened* (New York: St. Martin's Press, 2002), xii.

28 See Douglas Kellner, 'Media Spectacle and Media Events: Some Critical Reflections', in *Media Events in a Global Age*, eds. Nick Couldry, Andreas Hepp and Friedrich Krotz (London; New York: Routledge, 2010), 76–91.

29 Nicholas Mirzoeff, *An Introduction to Visual Culture* (London; New York: Routledge, 2009), 2. While acknowledging 9/11's media presence, Mirzoeff does not make a clear distinction between Western media and media generally. Although it is out of the scope of this chapter to further expand on this distinction, it is relevant to note that the media technologies to which Mirzoeff assumingly refers (such as the internet, digital video games, cinema and cameras) are based on specific circumstances for the US and Western (developed) world, and cannot be easily generalized as a rule. See, for example, "Most Afghans Don't Even Know About 9/11, According to Disturbing Poll," *Daily Mail*, 9 September, 2011, accessed November 7, 2017, http://www.dailymail.co.uk/news/article-2035160/Most-Afghans-know-9-11-according-disturbing-poll.html.

30 Mirzoeff, *An Introduction to Visual Culture*, 293.

31 See Ibid.

goes beyond US borders. More than anything else, it reveals the powerful connection this media event established with the global public.[32]

To understand the extent of the relationship between 9/11 and the media, and why this chapter considers 9/11 primarily as a media event, it is necessary to examine the distinction between 9/11 as an event and 9/11 as a media event. The 9/11 event affected the immediate family and relatives of the nearly 3,000 deceased, traumatized those who witnessed it and wrought enormous material damage to the Twin Towers site. If 9/11 is considered an event outside of the media, it can be seen to have influenced the WTC architectural space and to have assigned the site new socio-political and economic value after the destruction of the towers. As a media event, however, it first attracted the attention of a worldwide audience and produced a sense of fear, trauma, and sympathy and a globally visible iconotype. For the global public who witnessed it on television, this event, metaphorically speaking, 'existed' only in the media, which is why it is difficult to see 9/11 as separate from a media event.[33] The disaster marathons that occurred immediately after the event and the extensive tabloidization that followed in printed media, arguably turned this into media event.[34] This is why, while also considering that the event's media presence was one of the perpetrators' goals, the position in this chapter is to consider 9/11 predominantly as a media event.

The Media Image of the Twin Towers in Hindsight

Looking back on 9/11, a number of discourses from different disciplines positioned the architecture of the Twin Towers as crucial to understanding the

32 The global public is understood here as the Western public, mainly in North America and Europe.

33 Žižek adopts the same position in relation to 9/11, arguing that the general public's perspective, or 'our' perspective, is dependent on media officials, and that we should always be aware of this. See Žižek, *Desert of the Real*, 37. Additionally, Blondheim and Liebes argue that 9/11 is difficult to separate from its media presence, simply because so many have seen it in the media. Blondheim and Liebes believe that, in the case of 9/11, television coverage of the event became part of it, and outlined the difficulty of distinguishing between television coverage of the event and that same coverage being a part of the event. See Blondheim and Liebes, "Live Television's Disaster's Marathon," 274.

34 This is also supported by media theory, in which, as already mentioned, Kellner considers this media event a spectacle of terror; see, for example, Kellner, "Media Spectacle and Media Events," 76–91.

multiple phenomena which ensued after the event.[35] Architectural historian Andrew Herscher examines these in his insightful article on 9/11, 'The Language of Damage' (2002), explaining them as related to what he calls 'the discourse of the attack': 'as representation, the infliction of damage not only partially or totally destroys its target; it also marks its target in a certain way. What damage attempts to represent, according to what may be regarded as the discourse of the attack, is the relation of its target to the trauma that the attack responds to.'[36] Herscher continues that this damage also reveals architecture's potential to produce meaning after its destruction. He further holds that architecture is representational both in its construction and destruction. Considering the multiple discourses that have attempted to explain the causes and motives behind 9/11, in which the attack was either caused by Islamic terrorists or the Israeli secret service, or in which the US was either too militarily active towards its enemies or too tolerant towards them, Herscher asserts that violence and damage uncover both the ideological and non-ideological structures behind that violence. According to Herscher, in these attacks, victims, institutions, or nations develop symbolic relationships: 'victims become representatives of a targeted institution, nation, or phenomenon.'[37] Finally, Herscher concludes, although the ideological structures behind violence (on the part of both perpetrators and victims) are significant in these attacks, architecture remains both the context of such attacks and their core element.[38]

As further evidence that architecture of the Twin Towers is inextricably related to 9/11, it is relevant to note that the public, along with different disciplines (including architecture), looked for reasons why the Twin Towers were selected as targets in relation to their architecture, structure, function, or symbolism. Some debates invoked discourses of what can be called the 'latent target potential' of the towers, rendered thus in the aftermath of the attack. For example, philosopher Jean Baudrillard implicitly acknowledges this 'target potential' in his book *The Spirit of the Terrorism* (2002).[39] He wrote, 'at a pinch, we can say that they did it, but we wished for it ... Allergy to any definitive order, to any definitive power is—happily—universal, and the two towers of the World Trade Center were perfect embodiments, in their very twinness,

35 Achieving significance in hindsight is a crucial feature of the philosophical concept of event, as Žižek points out; see Slavoj Žižek, *Event: A Philosophical Journey through a Concept* (Brooklyn, London: Melville House, 2014), 4.

36 Herscher, "The Language of Damage," 70.

37 Ibid.

38 Ibid.

39 Jean Baudrillard, *The Spirit of Terrorism and Requiem for the Twin Towers* (London; New York: Verso, 2002).

FIGURE 3.1 *The World Trade Center and Battery park prior to 9/11, New York, US*
PHOTO: WIKIMEDIA COMMONS. CREDIT: PHOTOGRAPHS IN THE CAROL M. HIGHSMITH ARCHIVE, LIBRARY OF CONGRESS, PRINTS AND PHOTOGRAPHS DIVISION.

of that definitive order.'[40] For Baudrillard, 'order' denotes the Western hegemony of economic and political power, with the US as a core of this hegemony (figure 3.1).[41] Thereby, Baudrillard sees the *twin-ness* of the towers (or their architecture) as crucial to their target potential. Smith equally notes that it was the 'double value' of the target that was crucial to it being chosen, thus acknowledging the crucial role the Twin Towers had in this event.

Architectural historian Reinhold Martin also addresses the peculiar relationship between architecture and 9/11. He refers to Larry Silverstein, the developer who had just signed the 99-year lease of the World Trade Center before 9/11, and who expected to receive $3.6 billion in case of a catastrophic event. However, since these legal documents were still being finalized when 9/11 took place, Silverstein asked for additional clarification of the term 'event' in these documents. His belief was that he should receive twice the agreed sum, since

40 Ibid.
41 Id., 31–32.

two towers were destroyed, each hit by a different plane within approximately 15 minutes. Silverstein thereby argued that two events had taken place. Martin, on the contrary, concludes that the multiple planes and multiple attacks in 9/11 qualitatively resulted in *one event*. More than anything else, these debates acknowledge the relationship between Twin Tower's architecture and multiple post-9/11 debates.[42] These discourses are crucial for revealing the nuances of the post-9/11 media image of the Twin Towers, along with the peculiar positions and wider social, political, economic, and media contexts in which their architecture was positioned.

Martin also historically examines both the Twin Towers and the Pentagon, suggesting that both buildings, after 9/11, should be considered as historically intertwined. He argues that since 'both the structure and the symbolism of this event involve actual architectures, its historicity intersects with that of the buildings themselves.'[43] Martin notes that the Pentagon was completed in 1943, around the same period the Twin Towers were envisioned. Although the towers were not built until much later, a World Trade Corporation was established in 1946 by New York State to promote international trading of goods, a process that would rely on the Port of New York. This was followed by studies of the office space in buildings that were initially intended to be scattered around the city. In 1956, David Rockfeller, a Vice President of Chase Manhattan Bank and a member of one of the wealthiest families in the US, founded the Downtown Lower Manhattan Association (DLMA) to support his investment in the area. He initially invited Skidmore, Owings & Merril (SOM) architects to make schematic plans of the five to six million square meters of office space.[44] Rockefeller's initial idea was to build the new Chase Bank headquarters. In the end, he decided to construct new buildings as a part of an urban renewal in the area.[45] The Twin Towers project was, however, inherited and finally realized by the Port Authority of New York, the agency that overlooked the ports and their businesses in the city, and that had growing power in the 1960s.[46] Once under the Port Authority, the new project was given to Richard Adler in 1960 for further planning and study. Adler had the idea to design a skyscraper with more than 100 stories. In 1962, the public relations staff of the Port Authority were

42 Martin, "One or More," 115–117.

43 Id., 117.

44 Id., 118–119.

45 Deyan Sudjic, *The Edifice Complex: How the Rich and Powerful Shape the World* (London; New York: Allen Lane, 2005), 355.

46 The Port Authority, prior to its involvement in constructing the old WTC, took part in adding a new deck to the George Washington Bridge; it also played a part in the Lincoln Tunnel and Verrazano-Narrows Bridge construction projects. Smith, *The Architecture of Aftermath*, 103; Martin, "One or More," 119.

searching for an architect who could produce a more recognizable image for the project, and finally chose Minoru Yamasaki.[47] Emery Roth & Sons collaborated with Yamasaki on detailing and construction, and the plan to make the Twin Towers the tallest buildings in the world was revealed in 1964. Construction started in 1966—the first tower opened in 1973 and the second in 1976.[48]

The WTC was initially built to revitalize the local economy, which suffered after job losses in the shipping and manufacturing industry during the 1950s. Manhattan was an expensive port for transporting goods to and from New York, so shipping and travelling was often conducted via other, cheaper city ports, which were also more efficient. The idea was to turn a non-feasible area into a real estate business, which would invite foreign companies to move into the office complex and further trade with the US. The WTC was supposed to manage these export–import activities in one place, and represented hope for the city's better economic future.[49]

Although built with the intention of helping a declining city economy, the towers also created a new image of the city skyline that became synonymous with US political and economic power—power that was often assumed as such by those outside the US. Architectural historian Mark Wigley notes that, in terms of visual representation, the buildings were designed to attract global attention by rising 'up above the city at the end of the island facing Europe'.[50] Although the towers achieved symbolic value and soon became a tourist attraction, also appearing on postcards, architectural literature mostly criticized them, or even left them out of architectural history books.[51] This criticism often targeted the towers' generic architectural qualities, visible in the reductive simplicity and 'randomness' of their façades and the buildings' gigantic

47 Martin, "One or More," 118–120; for further information on the architect selection process, see David L. Salomon, "Divided Responsibilities: Minoru Yamasaki, Architectural Authorship, and the World Trade Center", *Grey Room* 7, Spring, Special issue: On 9/11 (2002): 86–95.

48 Martin, "One or More," 118–120.

49 Sudjic, *The Edifice Complex*, 355–356; Peter F. Cannavò, *The Working Landscape: Founding, Preservation, and the Politics of Place* (Cambridge: MIT Press, 2007), 127–130; Smith, *The Architecture of Aftermath*, 99–103, 104.

50 Mark Wigley, "Insecurity by Design", in *After the World Trade Center: Rethinking New York City*, eds. Michael Sorkin and Sharon Zukin (New York: Routledge, 2002), 73.

51 Smith noted that these skyscrapers were not included in the three editions David Watkin's *A History of Western Architecture*, published in 1986, 1996 and 2000. See David Watkin, *A History of Western Architecture* (London: Barrie & Jenkins, 1986); the towers were also left out of the 1986 and 2000 editions of *Architecture, from Prehistory to Post-Modernism: The Western Tradition*; see Marvin Trachtenberg and Isabelle Hyman, *Architecture, from Prehistory to Post-Modernism: The Western Tradition* (New York: H.N. Abrams, 1986).

FIGURE 3.2 *The Twin Towers, New York, US*
PHOTO: WIKIMEDIA COMMONS.

size (figure 3.2), along with their troubling relationship with Manhattan's urban grid. They represented the architectural language of modernism, and the architect firm chosen for the project, Minoru Yamasaki & Associates, was not one of the most highly regarded architecture firms of that time.[52] Architectural academia only saw genius in the twinness of the towers (largely after their destruction, however), which was initially envisioned by Yamasaki with the purpose of housing a sufficient amount of office space.[53]

An entire site was cleared for this project, which in the end consisted of seven buildings—the Twin Towers, four low plaza buildings and the Vista International Hotel.[54] The towers accommodated more than 430 businesses with around 50,000 employees. They included offices for banks, law firms, brokerage houses, television stations, publishers, charitable organizations, airlines, trade associations, representatives of foreign governments, and nine different chapels.[55] The towers were the tallest in the city and an iconic representation of what a city aspired to be: a mighty imperial center.[56]

52 Smith offers an extensive account of this criticism and general reactions to the towers in his book *The Architecture of Aftermath,* specifically on pages 111–117; some examples of this criticism include Manfredo Tafuri and Francesco Dal Co, *Modern Architecture*, 2 vols., History of World Architecture (New York: Electa/Rizzoli, 1986), 308.

53 Koolhaas and McGetrick, *Content*; Sudjic, *The Edifice Complex*.

54 Smith, *The Architecture of Aftermath*, 106.

55 "List of World Trade Center Tenants," *CNN*, accessed November 7, 2017, http://edition.cnn.com/SPECIALS/2001/trade.center/tenants1.html; see also "Companies in the Twin Towers: Before and After," *Fortune*, accessed November 7, 2017, http://archive.fortune.com/magazines/fortune/twintowers.

56 Sharon Zukin, "Our World Trade Center," in *After the World Trade Center: Rethinking New York City*, eds. Michael Sorkin and Sharon Zukin (New York: Routledge, 2002), 13.

Although some architectural literature saw the aftermath of 9/11 as an opportunity to reiterate prior criticism of the towers, architectural academia and architects agree on their image value. (This image value correlates directly with the towers being selected as targets; it arguably increased their target value.) Furthermore, the popularity of the buildings outside the field of architecture was arguably related to this image value. Wigley states that the towers were 'the centerpieces of billions of images'[57] and points of outlook towards the city for countless visitors. Considering that the majority of the New York public, along with tourists, were not architects or employed in the towers, Wigley argues that the buildings' presence over Manhattan was understood as 'the architecture of image' and was enjoyed by the general public.[58] Although they came behind the Empire State Building and Central Park in terms of visitor numbers, the towers defined the city skyline both physically and in the media. They made the skyline immediately recognizable; they were the tallest towers in the city and acted as orientation points for New York visitors and residents.[59] They were used as logos in advertisements and frequently appeared in movies.[60]

Furthermore, they were settings for dramatic and memorable stunts, such as multiple tightrope crossings by Philippe Petit in 1974, or two parachute jumps in 1974 and 1997.[61] This overall relationship with image is unsurprising, considering that the concept was part of Yamasaki's initial intention and the reason he was chosen as an architect. Smith also notes that the destroyed towers were 'key symbols within the later twentieth-century ... icons with the capacity to stand for crucial values ... the most visible point of concentration of the complex array of powers associated with them.'[62]

The Twin Towers were also designed to look representative in a 'pictorial still', since they were visible from planes and in photographs of the city skyline.[63] Architectural critic Sarah Williams Goldhagen proposes this term in reference to the Twin Towers, the Capitol Complex at Brasilia, designed by Oscar Neimeyer, and the Empire State Plaza in Albany, New York, designed by Wallace K Harrison:

---

57 Wigley, "Insecurity by Design," 73.

58 Id., 74.

59 Smith, *The Architecture of Aftermath*, 106.

60 Some of these movies are *Godspell* (1973), *King Kong* (1976) and *The Wiz* (1978).

61 Peter Skinner et al., *World Trade Center: The Giants Who Defied the Skies* (Vercelli, Italy: White Star, 2002), 23–25.

62 Smith, *The Architecture of Aftermath*, 6–7.

63 See Sarah Williams Goldhagen, "Monumentality in the Pictorial Still," in *Architecture between Spectacle and Use*, ed. Anthony Vidler (Williamstown; New Haven: Sterling and Francine Clark Art Institute; Distributed by Yale University Press, 2008), 89–106.

> These buildings were not principally designed with the people who would use them in mind ... what mattered to the men who built these complexes was how they looked in the pictorial still: the photograph, the view from the air, the view from the road. The goal was to make legible, imageable moments, and despite the reverberative onslaught of moving images in contemporary life from film and digital media, the pictorial still remains that image most likely to be committed to memory.[64]

Goldhagen thus acknowledges the importance of media (and image) for production of these skyscrapers, and essentially notes their ability to influence the creation of architectural space. Although she positions the towers outside the typical relationship between building and user, she notes the memory and image value of the Twin Towers, by which they communicate with the wider public. Considering that the towers were created with this image value in mind, their relationship with the public (and image) is key to their conception, and remains crucial in a post-destruction analysis. It further announces the public character of the new WTC; this chapter will now proceed to the analysis of this project.

The New World Trade Center

After 9/11, the site of the destroyed WTC became known as 'Ground Zero', a term that originates from Hiroshima and symbolizes ultimate destruction. New York's Ground Zero was a place of destruction and reconstruction, and, as Marita Sturken notes, determined by meanings tied to local, national and global discourses, and equally 'a neighbourhood, a commercial district, and a site of memory and mourning'.[65] The debris was removed after some time, leaving a void and giving the place a 'hallowed' or 'sacred' identity, owing to the number of victims whose remains will never be recovered (figure 3.3). Ground Zero was transformed from a place of emergency into one of tourism, media attention, and flows of people. During this transformation, there also arose an immediate need to assign new values to the site, which, as Sturken asserts, were different to those of fear, mourning, and nostalgia.[66]

Post-9/11 discourses were largely organized around concepts of 'rebuilding' and 'security'.[67] The rebuilding discourses questioned, for example, whether

64 Goldhagen, "Monumentality in the Pictorial Still," 100.

65 Marita Sturken is a professor in media, culture and communication. Sturken, "The Aesthetics of Absence," 311–312.

66 Sturken, "The Aesthetics of Absence," 314, 317, 319.

67 Martin, "One or More," 115.

FIGURE 3.3 *Ground Zero*
PHOTO: WIKIMEDIA COMMONS. CREDIT: NATIONAL OCEANIC AND ATMOSPHERIC ADMINISTRATION.

the towers should be rebuilt identically or differently, and how the deceased should be acknowledged at the site. In July 2002, Lower Manhattan Development Corporation (LMDC) and the Port Authority of New York and New Jersey organized the first round of an international architectural competition to design a new WTC. The finalists, which entered the second round of the competition in November 2002, were Studio Daniel Libeskind, Foster & Partners, Richard Meier & Partners (who designed their proposal as a part of the group with additional architectural firms), THINK Design and Architecture, Peterson/Littenberg Architecture and Urban Design, SOM (who also designed their proposal as a part of the group with additional architectural firms), and United Architects. The projects were judged on the basis of 12 criteria, which included price, public response, vision, connectivity, public space, and memorialization of 9/11 victims.[68] The project brief thus acknowledged the relevance of the public and their response in shaping the site's future architecture.

The architecture the finalists proposed largely relied on the symbolic representation of the WTC site previously associated with the Twin Towers and the image these new projects would project for both the site and the city, as

68 "World Trade Center Timeline of History," *World Trade Center*, accessed November 7, 2017, http://www.wtc.com/about/history.

well as inside and outside US borders.[69] Considering the unprecedented public attention given to this competition, the architects had the complex task of addressing the site's emotional and memory value, and producing the expected symbolic image value associated with the power representations of the previous Twin Towers. In doing so, as architectural historian Felicity D. Scott explains, the architects missed the opportunity to engage with shifting social, political, institutional, and historical contexts; this prompts Scott to question the future of architecture, along with the nature of its relationship with the present.[70] According to Scott, the projects were 'clichéd versions of social tropes', or elements with which the wider public and the media could easily relate with, such as heroes, monuments, rebirth, and more. Scott's analysis is relevant because it evidences the impact of the WTC reconstruction on architectural history.

In light of these circumstances, it is necessary to reflect on the new WTC's position as a public project, where the need to fill the gap created by the destroyed Twin Towers predominantly defined the design proposals. As Scott notes, this public attention caught the architectural discipline by surprise, and the competition finalists were not properly equipped to respond to the unconventional rebuilding circumstances. Scott rightfully explains the architects' disengagement with larger political discourses as the partial result of a general lack of interest in theory. While she further identifies the need to recover architectural attempts to critically engage with contemporary political, social, and technological circumstances, this is arguably only possible by fully engaging with the uniqueness of the circumstances around 9/11, including its media and image contexts. Furthermore, the notable shift from private to a project present in the public sphere, which necessarily maintains the relationship with image and contemporary media, represents an additional layer of this architecture that must be further examined. By acknowledging the nuances in the new WTC architecture's engagement with media, image, and the public, there is room to identify these phenomena and learn from these contemporary circumstances, and to proactively contribute to the nature of architecture in the public sphere in the future.

The public presence of this project was evident in the design competition process for the new WTC. In February 2003, Studio Daniel Libeskind's project was chosen. It proposed the tallest tower in the world, five smaller skyscrapers,

---

69 Karen Beckman, "Terrorism, Feminism, Sisters, and Twins: Building Relations in the Wake of the World Trade Center Attacks," *Grey Room* 7, Spring, Special issue: On 9/11 (2002), 30.

70 Felicity D. Scott, "Involuntary Prisoners of Architecture," *October* 106, Fall (2003), 75, 101; Felicity D. Scott, *Architecture or Techno-Utopia: Politics after Modernism* (Cambridge: MIT Press, 2007), 248, 280–281.

FIGURE 3.4 *Transportation Hub (left), WTC 3 (middle), and WTC 4 (right), 2017*
PHOTO: WIKIMEDIA COMMONS. CREDIT: ANTONY-22. LICENSE: CC BY-SA 4.0.

an elevated promenade circling the structure and a plaza called the Wedge of Light. The tallest tower was to be a symbolic 1,776 feet high, a reference to the year when the US Declaration of Independence was signed. Nevertheless, Libeskind's proposal very quickly attracted criticism.[71] In light of extensive public attention alongside criticism and political plays, Libeskind went from being the new WTC architect to its master planner. By July 2003, just a few months after Libeskind was announced as winner, it was decided that One WTC would be designed by David Childs of SOM. The project was altered, and a number of famous architects contributed by designing different buildings on the site (figure 3.4). It was redesigned by a group of architects including David Childs, Fumihiko Maki, Santiago Calatrava, Richard Rogers, Daniel Libeskind, Bjarke Ingels Group (BIG), and others.

71 Herbert Muschamp, "Balancing Reason and Emotion in Twin Towers Void," *The New York Times*, 6 February 2003, accessed November 7, 2017, http://www.nytimes.com/2003/02/06/arts/balancing-reason-and-emotion-in-twin-towers-void.html.

FIGURE 3.5 *WTC 1 (the tallest building; 2013)*
PHOTO: SANJA RODEŠ.

FIGURE 3.6 *WTC Memorial*
PHOTO: SANJA RODEŠ.

Libeskind winning the competition was largely a result of him appealing to public emotions; he gave a heartfelt speech on CNN Live about coming to the US as a migrant and his early fascination with the Statue of Liberty (he designed the tallest WTC tower which architecturally mimicked the statue).[72] In a further appeal to emotion, he also proposed the 'Park of Heroes', 'Wedge of Light', 'Gardens of the World', 'Memory Eternal Foundations' and similar as part of his design.[73] After Libeskind won, this crucial relationship with the public remained visible in the alterations to his project and how it was assigned to different architects, which was a response to the negative publicity his design received. Once completed, the new WTC will consist of One WTC (figure 3.5), WTC towers 2, 3, 4, 5 and 7, a National September 11 Memorial (figure 3.6) and Museum, the Transportation Hub, a Retail Space, and a Performing Arts Center. Tower 7 opened in 2006, the 9/11 Memorial in 2011, Tower 4 in 2013, the 9/11 Museum and One WTC opened in 2014, while the Transportation Hub opened in 2016.

9/11 as a 'Limit' of the New WTC

In order to examine the relationship between the new WTC and 9/11, the chapter turns towards Tschumi's concept of 'limit'. It argues that 9/11 is a limit of the new WTC architecture, and examines this concept as a form of translation. In his 1976–7 project, Tschumi published a number of posters named 'Advertisements for Architecture'. These were early theoretical texts, postcard-sized productions of images and words.[74] He argued that 'there is no way to perform architecture in a book. Words and drawings can only produce paper space, not the experience of real space. By definition, paper space is imaginary: it is an image.'[75] Tschumi emphasized that the majority of us experience architectural space when looking at printed media images (or, to be specific, in our imagination) rather than in reality.[76] His posters examined architecture as it relates to matters of political freedom, and, as Smith states, applying the most effective forms of public art for that moment in order 'to dramatize the underlying

72 Sudjic, *The Edifice Complex*, 365–367.

73 Scott, *Architecture or Techno-Utopia*, 268–269.

74 "Projects," *Bernard Tschumi Architects*, accessed November 7, 2017, http://www.tschumi.com/projects/19/#.

75 Ibid.

76 "Bernard Tschumi. Advertisements for Architecture," *WFW*, 7 December 2010, accessed November 7, 2017, http://www.we-find-wildness.com/2010/12/bernard-tschumi-advertisements-for-architecture/.

political currents of architecture itself'.[77] These posters examined the concerns to which architects at that time were oblivious.

Exploring the correlation between advertisements, media, and architecture, and specifically how images of buildings repeatedly presented in advertisements can create a desire to visit those buildings, Tschumi wrote for one poster: 'to really appreciate architecture, you may even need to commit a murder. Architecture is defined by the actions it witnessed as much by the enclosure of its walls. Murder in the Street differs from Murder in the Cathedral in the same way as love in the street differs from the Street of Love. Radically.'[78] The text was accompanied by the silhouette of a man falling from the window of a building. This well-known postcard seems 'both prescient and off the mark'[79] after 9/11. Tschumi stressed that architecture, along with other creative outputs and arts, sometimes finds itself with 'disturbing' elements at its limits. He emphasized that such limits are particularly valuable for defining architecture, which would be non-existent without them. He defined limit as

> a boundary ... Questioning limits is a means of determining the nature of discipline. Today's limits of architecture: (1) things pertaining to the relationship between spaces and their use, between "type" and "program", between objects and events; (2) things pertaining to the notation of architecture.[80]

According to Tschumi, architecture would not exist without its drawings or texts, along with the cultural, social, and philosophical demands that have developed and 'made architecture a form of knowledge in and of itself'.[81] He argues that these elements inform us about key architectural concepts more accurately than buildings, although they are not a built form of knowledge inherent to architecture. He named these elements the 'works of the limit',[82] arguing that these 'limits' are a key aspect that defines the content within the limit, where the content is architecture.

These statements announced different modes of thinking at the time they were published (the essays 'Architecture and Limits I, II, and III' appeared

77 Smith, *The Architecture of Aftermath*, 124.

78 "Projects."

79 Smith, *The Architecture of Aftermath*, 126.

80 Tschumi, *Architecture and Disjunction*, 101–102; Bernard Tschumi and Museum of Modern Art (New York), *The Manhattan Transcripts* (London; New York, NY: Academy Editions; Distributed to the trade in the USA by St. Martin's Press, 1994), xx.

81 Tschumi, *Architecture and Disjunction*, 102.

82 Tschumi, *Architecture and Disjunction*, 102.

FIGURE 3.7 *Public art installation 'Tribute in Light' which occurs yearly on 11 September, marking the anniversary of the 9/11 attack; intertwining of the Twin Towers' architecture, 9/11 and the new WTC*
PHOTO: WIKIMEDIA COMMONS. CREDIT: KIM CARPENTER. IMAGE: '9-11-11 WTC TRIBUTE IN LIGHT FROM JERSEY CITY, NJ'. LICENSE: CC BY 2.0.

in 1980 and 1981). Today, however, these definitions are well known in architectural theory. Nevertheless, it is important to reflect on Tschumi's 'limit' of architecture as another intuitive point (in addition to the poster discussed by Smith) that can offer a different perspective post-9/11, particularly on the (media) relationship between the Twin Towers, their destruction and the new WTC. Borrowing Tschumi's definition, one can argue that 9/11's act of terrorism presented such a 'limit' for the WTC architecture. Using Tschumi's proposal that the 'disturbing and out of character'[83] elements of architecture's limits are useful for defining its 'content', then the destruction of the Twin Towers on 9/11 can tell us more about their architecture. Not only this, but if the 'limits' of architecture define architecture, then the destruction of the Twin Towers directly influences the architecture of the new WTC (figure 3.7). This correlation represents a relevant paradigm in analyzing the relationship between this media event and the new WTC architecture, and further suggests ways in which 9/11 and both old and new WTC are interrelated.

9/11's media image influenced the new WTC's media presence by exposing it to the public prior to its existence. The new WTC is positioned within

83 Id., 101.

the contexts actively produced by such an event—contexts that are political, economic, social, and media-related. The 9/11 event, particularly its media presence, set the tone and contexts within which the new WTC designs were proposed, judged and further shaped. As related to the physical footprints on the site as a part of the 9/11 memorial, these footprints aimed to address the emotional and memory value associated with Ground Zero. If 9/11 is considered a limit of the new WTC, its media image and media presence are a necessary paradigm for further analyzing this architecture.

The CCTV and the Twin Towers Iconotype

The relationship between the CCTV and the new WTC can be argued to have been established in early 2002, when OMA received invitations to participate in two competitions: one was rebuilding the WTC, and the other was building the new CCTV Headquarters in Beijing. OMA decided not to enter the competition for the new WTC and chose to compete for the Beijing project instead, arguing that not much architecture could be created in the heavily politicized circumstances surrounding the WTC.[84] The previous CCTV consisted of an office tower and television complex, and OMA won the competition in 2002 to design a new building in the Beijing CBD. The new CCTV building is intentionally different to the old one, aiming to project China's power in the 21st century and strengthen Beijing's image in light of the 2008 Olympics.[85] The building was expected to be finished by 2008, but the neighboring Television Cultural Centre building (TVCC, also designed by OMA) caught fire in 2009, significantly delaying further construction. Following this, the new CCTV opened in 2012.[86]

---

84 Koolhaas and McGetrick, *Content*, 473, 483, 515. Sudjic argues that, even though Koolhaas claims that Asia is more vibrant and architecturally interesting than America, what actually made Koolhaas choose a different continent was 'a litany of disappointment'. Sudjic names a number of OMA projects that were abandoned or closed after opening: the plan for a new film studio for Universal in Hollywood was abandoned, along with the scheme to extend the Los Angeles 20-acre County Museum of Art (LACMA). The project of the extension of the Whitney Museum of American Art was aborted after three years of construction, while the Guggenheim in Las Vegas closed a year after opening because it had so few visitors. Sudjic, *The Edifice Complex*, 137.

85 Ying Zhu, *Two Billion Eyes: The Story of China Central Television* (New York, Jackson, TN: New Press; Distributed by Perseus Distribution, 2012), 8.

86 Andrew Jacobs, "China TV Network Apologizes for Fire," *The New York Times*, 10 February 2009, accessed November 7, 2017, http://www.nytimes.com/2009/02/11/world/asia/11beijing.html?_r=0.

Changing Image: The New Beijing CBD *and 'Building Big' in China*

Beijing's old architecture featured the old city with hutongs (narrow streets and alleys), which contrasted with the contemporary architecture in the city center. Beijing was growing in size in the late 20th and in 21st century and expanding concentrically, with new ring roads being constructed every few years.[87] The second ring road, which surrounded the city by 1980s, is now accompanied by the third, fourth, and fifth; the sixth opened in 2009, and the seventh will be finished by 2017.[88]

The fast growth of the cities and the construction boom in China is a product of China's economic growth during the 1990s, rather than of the deindustrialization process experienced in the West at the same time. This economic growth was a result of announced openness policy and economic reform after 1978. Between 1978 and 1988, the average annual growth of the Chinese economy was 9.5%.[89] The new architectural projects are a cumulative result of actions related to local investment capital and government support, along with designs delivered by international architectural firms. Unlike the projects in Western countries, the construction market was not influenced by global investment capital—instead, it was made using surplus from local capital accumulated in sectors such as mining, manufacturing, and export. As land prices increased towards the end of 20th century, real estate became an ideal sector to invest in, and the surplus of this capital was directed from the production sector towards construction and architecture.[90]

Moreover, government intervention in the construction boom was motivated by building image-producing projects, which would advance certain political careers and provide profit from land leasing.[91] As previously noted, the presence of the international architecture firms is a typical feature of China's construction boom. As urban sociologist Xuefei Ren argues, companies such as SOM or Foster & Partners, although very present internationally, were

---

87 Ole Scheeren, "Made in China," *Perspecta* 39, Special issue: Re_Urbanism: Transforming Capitals, eds. Tina Di Carlo et al., (2007), 39.

88 Sudjic, *The Edifice Complex*, 131; Xinhua, "Beijing's New Ring Road to Run 1,000 km," *China.org*, 7 July 2014, accessed November 7, 2017, http://www.china.org.cn/china/2014-07/07/content_32881371.htm.

89 Hans Ibelings and Powerhouse Company, *Shifts: Architecture after the 20th Century* (Amsterdam: Architectural Observer, 2012), 13; Shue Tuck Wong and Sun Sheng Han, "Whither China's Market Economy? The Case of Lijin Zhen," *Geographical Review* 88, no. 1 (1998): 29–46.

90 Xuefei Ren, "Architecture as Branding: Mega Project Developments in Beijing," *Built Environment* 34, no. 4, Special Issue: The Transition of Chinese Cities (2008), 520–521.

91 Ren, "Architecture as Branding," 520.

particularly visible in China. They have received strong political and financial support from local and central governments. The project for the National Theatre, located in the Tian'anmen Square in Beijing, was entrusted to French architect Paul Andreu in 1998 (the largest Chinese project given to a non-Chinese architect); mega-projects such as the Olympic Stadium, designed by Herzog & de Meuron and the CCTV by Koolhaas (figure 3.8), quickly followed, again with the strong political and financial support of central governments.[92] This emphasis on international architectural firms was explained by Xia Ming, scholar in international political economy, as the Chinese architectural profession needing to be competitive in terms of global architectural trends, and to develop a Chinese architectural style to match the country's fast development. This need for style is marked by the desire for development and need to overcome China's tradition of imitating ancient styles.[93]

The CCTV emerged as a part of the already-established practice of 'building big' in 20th century China. This practice had its beginnings in the Ten Great Buildings project—these were built in 1958–9, and had a total construction area of 639,000 m2—the project was argued to be part of the socialist legacy.[94] As early as the mid-20th century, the Great Hall of the People (one of the Ten Great Buildings) was designed to be Western neo-classical style, which was 'part of Jefferson's vision for the United States government, of Hitler's Third Reich and of Stalin's Soviet Union'.[95] Building big in China is thereby a result of interest in producing awe-inspiring buildings, a practice previously introduced

92 Ren, "Architecture as Branding," 520–521.

93 Xia Ming, "Guanyu Oulu Fengge De Sikao" [Thoughts on European Continental Style], *The Architectural Journal* 11 (1999): 44–45 (quotation translated by K. Sizheng Fan), quoted in K. Sizheng Fan, "Culture for Sale: Western Classical Architecture in China's Recent Building Boom," *Journal of Architectural Education* 63, no. 1, Special issue: Vernaculars in the Age of Digital Reproduction (2009), 72.

94 Tao Zhu, "Building Big, with No Regret," *Architectural Association School of Architecture* 63 (2011), 105, 108. Zhao Dongri, the director of the Technology Office of Beijing Urban Planning Bureau and one of the chief architects involved in the project, explained in 1960 that 'the building should be large, and its shape magnificent … The architectural style should be joyful and clear so as to become the symbol of our new era'. See Zhao Dongri, "On the Growth of a New Architecture Style through the Selection for the Schematic Design of the Great Hall of the People," Architectural Journal 2 (1960), 13–16.

95 Zhu, "Building Big," 106. K. Sizheng Fan also argues similarly—that neo-classical style was a direct influence of Soviet social ideology, in which architects were encouraged to use classical architecture to achieve socialist monumentality. This practice also influenced China's architecture in the first decade of the People's Republic (1949–59). See Fan, "Culture for Sale," 64.

FIGURE 3.8 *CCTV building in Beijing, China*
PHOTO: WIKIMEDIA COMMONS. CREDIT: VERDGRIS. IMAGE: 'CCTV BUILDING IN BEIJING', NO CHANGES MADE. LICENSE: CC BY-SA 3.0.

by Hitler, Stalin, and Mussolini. The practice of building big has existed in China for more than 2,000 years, considering that all great dynasties would destroy the architecture of previous emperors and build new, representative buildings as a part of their political legacy.[96] Architectural scholar Tao Zhu emphasizes that, from the 1990s onwards, nationalism continues to serve as the dominant ideological foundation for a major portion of China's architectural output. This is especially apparent in the government's obsessive reliance on grand architectural projects and spectacular events, such as the Beijing Olympics, as a way to gloss over social contradictions, boost national pride and 'save face'—in other words, gain the respect and approbation of the outside world.[97]

Internationally famous architects have been used to promote the projects of local governments and developers internationally to investors and tourists. These were also tools to change Beijing's image from that of a post-socialist city

96 Zhu, "Building Big."

97 Id., 109.

to an international, architecturally relevant metropolis. After China entered the World Trade Organization in 2001, the service sector opened to international design firms, which largely contributed to the construction boom and the process of globalizing Chinese cities.[98] Beijing had a reputation for being the political and cultural center of China in the late 20th century when the government made a decision to make it the major financial center as well.[99] This decision was a result the competition among Chinese cities, and in order to be the financial center, Beijing had to compete with the Chinese financial capital—Shanghai. The decision to build Beijing's new CBD was also a result of this competition—the new CBD was positioned against Shanghai's financial district, Pudong, which was built in the 1990s. The plan to create a CBD was proposed in the Beijing General City Plan, which was approved in 1993 by the State Council. In 1998, the Beijing Government also provided the Specific Controlling Plan of Beijing, proposing the Chaoyang suburb as the new CBD location.[100] According to this plan, the CBD would house 'business[es], exhibition[s], hotel[s], residence[s] and entertainment function[s] and would become a new headquarters and management centre for financial, insurance, telecom, and information companies.'[101] The new district, envisioned in an industrial and residential area, resulted in the relocation of industrial facilities and the demolition of residential buildings, creating the tabula rasa conditions for the new architecture and the CBD.

The master plan for the CBD was a result of an international competition held in 2000, for which eight entries were selected (out of which only two were Chinese) to further submit their designs. The accent on internationalism in the selection process was obvious—the jury comprised Chinese and international experts. The entire competition process was mediatized, portraying the international companies' presence as a great success, with the aim of marketing the new CBD.[102] The city government emphasized that skyscrapers would be the major typology in the CBD, with projects such as the CCTV, World Trade Tower, Silktie Center, and Jianwai SOHO, some of which expected to be more than 300 meters high. These now form the symbolic building block of the CBD, symbolizing the new Beijing as a global city, with architecture that aims, as Ren writes,

98 Ren, "Architecture as Branding," 521.

99 The city also has a reputation of being a media capital, as argued in 2011 by Joshua Neves. See Joshua Neves, "Beijing En Abyme: Outside Television in the Olympic Era," *Social Text 107* 29, no. 2 (2011), 24.

100 Ren, "Architecture as Branding," 521–522.

101 Id., 522–523.

102 Id., 523.

to 'shock and impress' with concentrated skyscrapers around major highways through the CBD (figure 3.9). The government issued a number of favorable policies to attract companies to the CBD, which successfully resulted in a number of multinational companies moving to the area by 2004.[103]

Urban sociologist Broudehoux emphasizes that spectacular architecture has a central role in 'spectacularizing' the urban environment, and at the same time representing a 'smokescreen to hide the shortcomings of China's breakneck marketization, accompanied by rampant land speculation, corruption,

FIGURE 3.9 *CCTV building in Beijing, China*
PHOTO: WIKIMEDIA COMMONS. CREDIT: VMENKOV. IMAGE: 'VM BEIJING 3RD RING ROAD AND CCTV HEADQUARTERS 4516', IMAGE COLOR/BRIGHTNESS ALTERED. LICENSE: GNU FREE DOCUMENTATION LICENSE AND CC BY-SA 3.0.

103 Id., 523–524. Also of interest for comparison with the New York WTC site is that China's World Trade Center project was constructed in 1985–2010, also in Beijing's CBD: the 'international Beijing Central Business District is planned to be a central area of international business, and the goal of planning is to build CBD as the next World Trade Center in terms of scale and function'. See Beijing Central Business District Administrative Committee, "The Main Characteristics of Beijing Central Business District Comprehensive Planning Program," *Beijing Central Business District*, 11 June 2009, accessed November 7, 2017, http://203.86.47.237/en/construc/master/5825382375858826941.shtml.

rising inequities, and sociospatial polarization.'[104] While the poor have been evicted from their land and that land sold to private developers, the construction of megaprojects continues to boom in order to project the new city image. Broudehoux, however, goes one step further arguing that positive change occurred—Beijing and its government's actions were highly visible in the media, particularly in regard to the Olympics, in which the local architects, academics, intellectuals, and others saw an opportunity to react by protesting the corruption, inequalities, and land speculations, and demanding changes. These reactions resulted in budget and design modifications, and even the Beijing Government, in light of new media attention, was forced to operate in a more transparent manner.[105] The media presence and visibility thereby were productive for creating the opportunity for Beijing's residents to partially participate in its transformation.

Going East

The construction boom in Asia resulted in a cumulative number of high-rises that surpassed the number of existing skyscrapers in North America.[106] Koolhaas uses the expression 'Go East'[107] in his book *Content* (2004) to argue that architecture is 'moving' (or should move) to another continent, thus justifying his personal ambitions for the Asian market. He explains these ambitions with 'the fall of the wall, the new gravity of Eastern Europe, the emergence of China, the disarray of Japan', for which he further argues that these 'launched eastward momentum, accelerated by 9-11 and America's new preoccupations.'[108] When the competition for the CCTV project launched in 2002, Beijing was chosen to host the 2008 Olympic Games, which resulted in a construction completion date being set for the year of the Olympics.[109] Since the decision to build the CCTV was made ahead of the Olympics, it was intended as one of the city's

104 Anne-Marie Broudehoux, "Images of Power: Architectures of the Integrated Spectacle at the Beijing Olympics," *Journal of Architectural Education* 63, no. 2, Special issue: Changing Asia (2010), 61.

105 Moreover, young Chinese architects have formed groups that openly oppose spectacular brand-architecture, producing locally inspired urban designs and attempting to change built environment, along with Chinese architectural identity, on a smaller scale. For additional analysis, see Broudehoux, "Images of Power," 59–61.

106 Koolhaas and McGetrick, *Content*, 470.

107 Id., 104–508.

108 Id., 104.

109 Schereen, "Made in China," 45.

image-projects. The intention was 'to circulate as an image of power while China occupied the world stage as a host of the Olympics.'[110] The CCTV became an immediate icon for the international architectural audience.[111] This was due to its outstanding shape and size, and by the instantaneous attention that OMA project attract in the architectural media. Although an architectural icon, the CCTV is equally one of the most powerful propaganda systems in China under the Chinese Communist Party (CCP) Central Propaganda Ministry, which has a monopoly over TV broadcasting and is also profit-driven.[112]

The CCTV project, once compared with the new WTC, emerged from, of course, significantly different urban and socio-political contexts. Despite this, the correlation between the Twin Towers and the CCTV was established by the architect of the latter building, considering that the CCTV is based on the joined twin towers. Koolhaas' fascination with 9/11 and these towers, particularly their twinness, is evident in *Content*.[113] In explaining the concept of the CCTV in his 'Patent for "Bent Skyscraper"', Koolhaas wrote: 'by breaking the traditional tower that merely goes up and down at four points, a loop of building can be generated that unites and confronts its population in a single whole and cements a coherence of elements, isolating and separating them.'[114] As seen in this 'patent', Koolhaas presents the new CCTV as an embodiment of OMA's enthusiasm for experimenting with skyscraper typology. He presents this as necessary in his criticism of the skyscraper, which 'has not been refined,

110 Craig Johnson, "Icon and Ideology", in *Architecture, Disciplinarity, and the Arts*, eds. Andrew Leach and John Macarthur (Ghent: A&S Books, 2009), 174; Kayoko Ota, "Toyo Ito: Big Time Dilemmas," in Rem Koolhaas and Brendan McGetrick, *Content: Triumph of Realization* (Köln: Taschen, 2004), 448–449.

111 An architectural audience, or public, is conceivably defined as a part of architectural culture, such as architects and those with some education in architecture and its cognate fields, while a general public is understood as consisting of average media consumers (where media is television, internet or printed media). This definition, of course, admits the ambiguity of any borders drawn to separate one hypothetic group from the other, emphasizing that these groups are convenient constructs rather than firm categories.

112 Zhu, "Building Big," 110; see also Richard Lacayo, "The Architecture of Autocracy," *Foreign Policy* 166, May–June (2008).

113 Rem Koolhaas, "White Briefs Against Filth. The Waning Power of New York," in Rem Koolhaas and Brendan McGetrick, *Content: Triumph of Realization* (Köln: Taschen, 2004), 237, 511.

114 Koolhaas and McGetrick, *Content*, 511. In addition to this parallel, Koolhaas emphasizes that two of the engineers included in the post-9/11 reconstructions were also involved in the CCTV project. See Koolhaas and McGetrick, *Content*, 514.

but corrupted; the promise it once held ... has been negated by repetitive banality.'[115] The circular loop proposed for the CCTV is the most obvious attempt to subvert the idea of a skyscraper striving solely towards heights; Koolhaas imposes a skyscraper of infinite 'length', announcing the ultimate victory in the endless competition to create the tallest structure.[116] The joined towers of the CCTV are 234 m and 194 m high, with 540,000 m2 of gross floor area.[117]

As previously noted, Koolhaas presents the discourse around the CCTV's creation as related to the discourse of the Twin Towers' destruction. This is evident along two lines. First is the argument around the metaphorical 'killing' of the traditional skyscraper typology, while the second proposes the necessity of experimenting with skyscraper typology. The first line is obvious in the 'bleeding', 'stabbed' high-rise photomontage in *Content*, which aggressively aims to catch the reader's attention.[118] It is also identifiable in one of Koolhaas' statements on the CCTV, where he says, 'I want to kill the traditional idea of the skyscraper—it has run out of energy.'[119] Furthermore, he describes the Twin Towers as an example of the last 'invention' in the skyscraper.[120] In the second line of his argument, he presents the CCTV as OMA's vision of this new skyscraper typology.[121]

In relation to this invention in skyscraper typology, images of the CCTV project were an important part of the *Progress* exhibition, which was curated by architectural practice Rotor in 2011 and 2012 at the Barbican Gallery in London, and exhibits OMA projects and material.[122] The exhibition focused on exposing

115 Rem Koolhaas, "Kill the Skyscraper: Beijing CBD Core," in Rem Koolhaas and Brendan McGetrick, *Content: Triumph of Realization* (Köln: Taschen, 2004), 473.

116 "CCTV—Headquarters, China, Beijing, 2002. New Headquarters for China Central Television." *OMA*, accessed November 7, 2017, http://oma.eu/projects/cctv-headquarters.

117 "China Central Television (CCTV) Headquarters," *Arup*, accessed November 7, 2017, http://www.arup.com/projects/china_central_television_headquarters.aspx.

118 Koolhaas and McGetrick, *Content*, 473, 508.

119 Sudjic, *The Edifice Complex*, 140.

120 Koolhaas and McGetrick, *Content*, 443, 473.

121 The inspiration of the Twin Towers for the CCTV is also noted by Sudjic, who quoted Cecil Balmond, an engineer from Arup, a firm that contributed to the CCTV's design and construction: 'cut the loop, unfold the pieces, and you end up with the Twin Towers of the World Trade Center.' Sudjic, *The Edifice Complex*, 134; see also Johnson, "Icon and Ideology," 172.

122 Rowan Moore, "OMA/Progress Review," *The Guardian*, 9 October 2011, accessed November 7, 2017, https://www.theguardian.com/artanddesign/2011/oct/09/rem-koolhaas-oma-progress-barbican; "OMA/Progress," *OMA*, accessed November 7, 2017, http://oma.eu/projects/oma-progress.

OMA projects in a realistic environment, showing correspondence between investors, or discarded models alongside the final ones. While exhibiting photography, models, films, and other OMA's material, their work was presented in a 'messy', working environment to expose their production process from the perspective of different architectural practice (Rotor).[123] Material related to the CCTV was one of the highlights (considering that the project was completed a year after the exhibition opened). While the exhibition was called *Progress* and officially denoted 'work in progress', it also suggests broader matters of architectural progress and the critical relationship OMA maintains with the architectural present. The exhibition 'communicates ... [a] postmodern notion of time: the eruption of ideas explored through the designs recur, evolve or are simply dismissed; sometimes sustained by extreme rigour, sometimes relishing the unpredictability of intuitive thinking.'[124] The CCTV concept is timely and arguably related to the exhibition's critical position—Koolhaas sees the CCTV as an embodiment of invention in skyscraper typology, and thereby implicitly questions the matter of progress in skyscraper architecture.

Historian Kenneth Frampton notes Koolhaas' aspiration towards the invention of the skyscraper, and compares the CCTV with architect El Lissitzky's Wolkenbügel (meaning 'cloud-hanger') skyscraper proposal back in 1924. Frampton asserts that both CCTV and Wolkenbügel skyscrapers are equally technologically daring. More importantly, he also observes that the theoretical discourse of reinvention and progress argued by Koolhaas resembles El Lissitzky's concept of the 'anti skyscraper' in this Wolkenbügel proposal of horizontal skyscrapers (figure 3.10). El Lissitzky was a part of the Asnova (Association of New Architects) group established in 1923, which had as one of its goals 'a more scientific aesthetic',[125] along with exploring new architectural forms that would align with the terms of the new socialist state. As Frampton argues, Lissitzky attempted to 'reconstitute the American skyscraper in a socialist form', which translates as very close to the Koolhaas' theory of reinventing the skyscraper, where he conceptually 'recycled' the Twin Towers into a new skyscraper for Beijing's socio-political context. The reinvention of the skyscraper also relates to an investigation of Chinese architectural identity, which is supported by the expanding of architectural markets from the West to other parts of the

123 "OMA/Progress Review"; "OMA/Progress."

124 Roberto Bottazzi, "Casting a Critical Eye over OMA's Oeuvre," *The Architectural Review*, 31 October 2011, accessed November 7, 2017, https://www.architectural-review.com/rethink/reviews/casting-a-critical-eye-over-omas-oeuvre/8621652.article.

125 Kenneth Frampton, *Modern Architecture: A Critical History* (London; New York: Thames & Hudson, 2007), 345, 172.

FIGURE 3.10 *El Lissitzky's Wolkenbügel skyscraper proposal in 1924; photomontage*
PHOTO: WIKIMEDIA COMMONS. IMAGE IN PUBLIC DOMAIN [PD-US, PD-AUSTRALIA].

world.[126] Thereby, the discourse around 9/11 represented, to Koolhaas, a trigger suggesting the need for change and reinvention. Frampton further describes Lissitzky's skyscraper as 'an elevated propylaeum opening onto the peripheral boulevard surrounding the centre of Moscow. This work ... was intended as a critical antithesis to both the capitalist skyscraper and the Classical gate.'[127] This criticism parallels Koolhaas' critical position on skyscraper typology. The ambitious intentions of the architect are evident in terms of the daring skyscraper design and in the attempt to examine Chinese architectural identity and skyscraper typology.

The CCTV and WTC: The Relevance of Image-Architecture for Contemporary Architecture

The crucial significance of the CCTV architecture is for its experimentation with image architecture, and its subsequent position within wider architectural

126 Rem Koolhaas and Peter Eisenman, "Urgency 2007: Rem Koolhaas and Peter Eisenman," lecture, *CCA*, 8 June 2007, accessed November 7, 2017, http://www.cca.qc.ca/en/events/2681/urgency-2007-rem-koolhaas-and-peter-eisenman.

127 Frampton, *Modern Architecture*, 172.

discourses. The CCTV largely implicates different media images, both in the level of its conception (as related to the 9/11 iconotype) and in an attempt to project an image of China's economic power in the public sphere.[128] The CCTV and a number of other projects by OMA remain relevant examples of architecture-image relationships, and should be productively examined as such.[129] As architectural theorist and critic Aaron Betsky notes, Koolhaas maintains 'the pose of the architect as mythic maker, manipulator and mirror of data-driven imagery.'[130] While noting the seductive power of his architecture, and his focus on producing recognizable forms, Betsky concludes that 'it is not easy to dismiss Rem Koolhaas', and that he is able 'to produce a convincing architecture of image'.[131] Research into what exactly the 'convincing architecture of image' is represents the key to productively analyzing contemporary architecture in the eyes of media, and is further crucial to contemporary architectural discourses. The necessity of expanding the 'zone' of convincing image-architecture, and the field of productive analysis into these, remains critical in the present architectural environment, which has new and complex relationships with image and media. Without the intention of limiting such analyses to a one-architect examination, it is, however, relevant to emphasize that OMA's work can offer fruitful examples for this analysis.[132]

If media event is able to produce architecturally related iconotypes, and such iconotype is easily translated into other built examples by the mechanisms of globalization and global media, it is relevant to emphasize that media and image thereby necessarily enter architectural discourses. Considering that the CCTV was made to support Beijing's new image for the 2008 Olympics, the building is thereby correlated with both media events: the Olympics and 9/11. Rather than examining the physical transformation that mega events such as the Olympics can bring to an urban environment, this chapter outlines the

128 Johnson, "Icon and Ideology," 174.

129 For some examples of this analysis, see Hal Foster, *The Art–Architecture Complex* (London: Verso, 2011), 12–13; Aaron Betsky, "Rem Koolhaas: The Fire of Manhattanism Inside the Iceberg of Modernism," in Rem Koolhaas et al., *Considering Rem Koolhaas, and the Office for Metropolitan Architecture: What Is Oma*, ed. *Véronique Patteeuw* (Rotterdam: NAiPublishers, 2003), 39.

130 Betsky, "Rem Koolhaas," 25.

131 Id., 38–39.

132 Some of the Koolhaas' projects were defined already as able to productively and critically engage with wider political discourses. Scott identified one of these projects in her analysis of the entries for the new WTC contrasting this with an analysis of Koolhaas' thesis, *Exodus, or the Voluntary Prisoners of Architecture: The Strip* (1972), done at London's Architectural Association, in collaboration with Elia Zenghelis, Madelon Vriesendorp and Zoe Zenghelis. See Scott, *Architecture or Techno-Utopia*, 253–261.

more subtle impacts of media events.[133] These subtle impacts are established via media images of these events, and are thus part of the iconomy and global imaginary of the Western architect. Therefore, highly visible projects such as the CCTV mirror the media reality in architecture, in both its local and global contexts, where the most visible media events can cumulatively influence new and spectacular projects.

The new WTC and CCTV reveal the image-related processes in the production of this architecture. In both projects, media and image remain one site of their production, whereby both projects were produced with the expectation of being representative in previously mentioned 'pictorial still'. Furthermore, these examples effectively represent a contrast in their demonstrated ability to critically engage their architecture with wider political, social, or economic contexts. While Scott, as previously mentioned, criticized entries to the new WTC architectural competition, the CCTV represents a critical attempt to question the image-architecture, and to reinvent post-9/11 skyscraper typology. Without proposing the uniform 'solution' to image-architecture relationships, this chapter rather acknowledges the multifaceted character of the analysis into these.

In the public sphere, images of architecture and the urban environment are often used as a representational tool for communicating particular public messages and public image. For these reasons, this chapter argues that the relationships between architecture, image, and media, once in the public sphere, cannot be reduced to the dismissive analysis of architecture's engagement with image via concepts that have negative connotations, such as 'starchitecture' or spectacle architecture.[134] This chapter aims to expand the

133 For further information on the relationship between mega events (particularly the 2008 Olympics) and the urban environment, see Harry H. Hiller, "Toward an Urban Sociology of Mega-Events," in *Constructions of Urban* Space. Research in Urban Sociology, Volume 5, ed. Ray Hutchison (Bingley, UK: Emerald Group Publishing, 1999), 187; Broudehoux, "Images of Power"; Jingxiang Zhang and Fulong Wu, "Mega-Event Marketing and Urban Growth Coalitions: A Case Study of Nanjing Olympic New Town," *The Town Planning Review* 79, no. 2/3, Special issue: Planning the Chinese City (2008); Even Smith Wergeland, "When Icons Crumble—the Troubled Legacy of Olympic Design," *Journal of Design History* 25, no. 3, Special issue: Design Histories of the Olympic Games (2012); Megan Evans, "'Brand China' on the World Stage: Jingju, the Olympics, and Globalization," *TDR (1988-)* 56, no. 2 (2012).

134 See, for example, Hal Foster, *Design and Crime: And Other Diatribes* (London: Verso, 2002), 15–50; Foster, *The Art–Architecture Complex*. For further definitions and analysis of spectacle, see Guy Debord, *Society of the Spectacle*, 1 vols. (Detroit: Black & Red, 1983), and Guy

field and understanding of image architecture, demonstrating that these relationships can be productive for architecture, architectural images, and architectural theoretical discourses. The chapter asserts that analyzing this architecture has to account for the complexity of the relationships it maintains with image and media once in the public sphere. It emphasizes the necessity of continuously negotiating the position architecture occupies in these contexts, arguing that this allows for a better understanding of this architecture and its further development.

Regarding the significance of the media and image presence, both the new WTC and the CCTV act as global nodes of influence, and further represent the assertion of political will, as projects produced to create an expected image.[135] This global node is defined by architecture, which is highly present in media and represents an iconotype; it is also an architectural point of interest, a tourist spot, and an economic and socio-political representation of power. As global nodes, these projects thus mirror the significance of their economic and socio-political circumstances. As witnessed in these projects, their architecture maintains relationship with media image in an increasingly similar manner, despite different physical locations and differences between East and West.

The importance of reflecting on 9/11's impact on architecture and the translation of its media image lies in exposing the potential of a media event to assign new values to a particular space, and to influence the creation of architecture. If the impact of this media event and its image on architecture defined a limit of the new WTC, and further influenced the creation of the CCTV, then media event (in this instance, 9/11) remains a relevant framework within which these projects are considered. Media events and architectural media images are thereby able to influence architectural projects, and are often crucial in producing architecture with ambitious national importance. Thereby, it is relevant to establish that analyzing the relationship between architecture and media must occupy additional space in contemporary architectural analysis. While contemporary architectural literature often questions the nature of the discipline in the future, this type of inclusive analysis into architecture, image, and media can offer productive lines of further research, and provide architects with the tools to critically engage with contemporary architecture.

Debord, *Comments on the Society of the Spectacle*, trans. Malcolm Imrie (London; New York: Verso, 1998).

135 In the case of the new WTC, 9/11 received significant media attention, and the site of the new WTC was defined by a proliferation of images of the Twin Towers' collapse, which helped create this global node of influence.

Bibliography

Alexander, Michele G., Shana Levin, and P.J. Henry. "Image Theory, Social Identity, and Social Dominance: Structural Characteristics and Individual Motives Underlying International Images." *Political Psychology* 26, no. 1 (2005): 27–45.

Baudrillard, Jean. *For a Critique of the Political Economy of the Sign*. St. Louis, Mo.: Telos Press, 1981.

Baudrillard, Jean. *Simulacra and Simulation*. Translated by Sheila Faria Glaser. Series The Body in Theory. Ann Arbor: University of Michigan Press, 1994.

Baudrillard, Jean. *The Spirit of Terrorism and Requiem for the Twin Towers*. London; New York: Verso, 2002.

Baudrillard, Jean. *The System of Objects*. London; New York: Verso, 1996.

Beckman, Karen. "Terrorism, Feminism, Sisters, and Twins: Building Relations in the Wake of the World Trade Center Attacks." *Grey Room* 7, Spring, Special issue: On 9/11 (2002): 24–39.

Beijing Central Business District Administrative Committee. "The Main Characteristics of Beijing Central Business District Comprehensive Planning Program." *Beijing Central Business District*. Accessed November 7, 2017. http://203.86.47.237/en/construc/master/5825382375858826941.shtml.

Benjamin, Walter and Hannah Arendt. *Illuminations*. Translated by Harry Zohn. London: Fontana Press, 1992.

Best, Steven, and Douglas Kellner. "Debord, Cybersituations, and the Interactive Spectacle." *SubStance* 28, no. 3, Special Issue: Guy Debord (1999): 129–156.

Betsky, Aaron. "Rem Koolhaas: The Fire of Manhattanism Inside the Iceberg of Modernism." In Rem Koolhaas et al., *Considering Rem Koolhaas, and the Office for Metropolitan Architecture: What Is Oma*. Edited by Véronique *Patteeuw*, 25–39. Rotterdam: NAiPublishers, 2003.

Blondheim, Menahem and Tamar Liebes. "Live Television's Disaster's Marathon of September 11 and Its Subversive Potential." *Prometheus* 20, no. 3 (2002): 271–276.

Bottazzi, Roberto. "Casting a Critical Eye over OMA's Oeuvre." *The Architectural Review*. 31 October 2011. Accessed November 7, 2017. https://www.architectural-review.com/rethink/reviews/casting-a-critical-eye-over-omas-oeuvre/8621652.article.

Broudehoux, Anne-Marie. "Images of Power: Architectures of the Integrated Spectacle at the Beijing Olympics." *Journal of Architectural Education* 63, no. 2, Special issue: Changing Asia (2010): 52–62.

Cannavò, Peter F. *The Working Landscape: Founding, Preservation, and the Politics of Place*. Cambridge: MIT Press, 2007.

Carey, James W. *Communication as Culture: Essays on Media and Society*. Series Media and Popular Culture. London Boston: Unwin Hyman, 1988.

"CCTV—Headquarters, China, Beijing, 2002. New Headquarters for China Central Television." *OMA*. Accessed November 7, 2017. http://oma.eu/projects/cctv-headquarters.

"China Central Television (CCTV) Headquarters." *Arup*. Accessed November 7, 2017. http://www.arup.com/projects/china_central_television_headquarters.aspx.

Chun, Wendy Hui Kyong and Thomas Keenan. New *Media, Old Media: A History and Theory Reader*. New York: Routledge, 2006.

Colomina, Beatriz. *Privacy and Publicity: Modern Architecture as Mass Media*. Cambridge: mit Press, 1994.

Colomina, Beatriz and Joan Ockman. *Architectureproduction*. Revisions, Papers on Architectural Theory and Criticism. New York, NY: Princeton Architectural Press, 1988.

"Companies in the Twin Towers: Before and After." *Fortune*. Accessed November 7, 2017, http://archive.fortune.com/magazines/fortune/twintowers.

Cottle, Simon. *Media Organisation and Production*. Media in Focus. London; Thousand Oaks, Calif.: Sage, 2003.

Cottle, Simon. "Mediatized Rituals: Beyond Manufacturing Consent." *Media, Culture & Society* 28, no. 3 (2006): 411–432.

Couldry, Nick. *Media Rituals: A Critical Approach*. London; New York: Routledge, 2003.

Couldry, Nick, Andreas Hepp, and Friedrich Krotz, eds. *Media Events in a Global Age*. London; New York: Routledge, 2010.

Crysler, C. Greig, Stephen Cairns, and Hilde Heynen. *The Sage Handbook of Architectural Theory*. London: SAGE, 2012.

Dayan, Daniel and Elihu Katz. *Media Events: The Live Broadcasting of History*. Cambridge: Harvard University Press, 1992.

Debord, Guy. *La Societe Du Spectacle*. Paris: Buchet/Chastel, 1967.

Debord, Guy. *The Society of the Spectacle*. Translated by Donald Nicholson-Smith. New York: Zone Books, 1994.

Debord, Guy. *Comments on the Society of the Spectacle*. Translated by Malcolm Imrie. London; New York: Verso, 1998.

"Design Overview." *9/11 Memorial*. Accessed November 7, 2017. https://www.911memorial.org/design-overview.

Di, Carlo, Cao Fei Tina, Sze Tsung Leong, Ou Ning, Ole Scheeren, Yan Lei, and Zhang Jinli. "Texting Beijing." *Perspecta, Re_Urbanism: Transforming Capitals* 39 (2007): 30–47.

Dongri, Zhao. "On the Growth of a New Architecture Style through the Selection for the Schematic Design of the Great Hall of the People." *Architectural Journal*, no. 2 (1960): 13–16.

Dupré, Judith. *Skyscrapers*. New York: Black Dog & Leventhal, 2008.

Easterling, Keller. "Enduring Innocence." *Grey Room* 7, Spring, Special issue: On 9/11 (2002): 106–113.

Evans, Megan. "'Brand China' on the World Stage: Jingju, the Olympics, and Globalization." *TDR (1988-)* 56, no. 2 (2012): 113–130.

Fan, K. Sizheng "Culture for Sale: Western Classical Architecture in China's Recent Building Boom." *Journal of Architectural Education* 63, no. 1, Special issue: Vernaculars in the Age of Digital Reproduction (2009): 64–74.

Foster, Hal. *Design and Crime: And Other Diatribes*. London: Verso, 2002.

Foster, Hal. *The Art–Architecture Complex*. London: Verso, 2011.

Frampton, Kenneth. *Modern Architecture: A Critical History*. London; New York: Thames & Hudson, 2007.

Goldberger, Paul. *The City Observed, New York: A Guide to the Architecture of Manhattan*. 1st edition. New York: Vintage, 1979.

Goldhagen Williams, Sarah. "Monumentality in the Pictorial Still." In *Architecture between Spectacle and Use*, edited by Anthony Vidler, 89–106. Williamstown; New Haven: Sterling and Francine Clark Art Institute; Distributed by Yale University Press, 2008.

Habermas, Jürgen. *Strukturwandel Der Öffentlichkeit: Untersuchungen Zu Einer Kategorie Der Bürgerlichen Gesellschaft*. Politica. Neuwied, Berlin: Luchterhand, 1962.

Habermas, Jürgen. *The Structural Transformation of the Public Sphere: An Inquiry into a Category of Bourgeois Society*. Studies in Contemporary German Social Thought. Cambridge: MIT Press, 1989.

Habermas, Jürgen, Jacques Derrida, and Giovanna Borradori. *Philosophy in a Time of Terror: Dialogues with Jürgen Habermas and Jacques Derrida*. Chicago: University of Chicago Press, 2003.

Hallin, Daniel C., and Paolo Mancini. "Summits and the Constitution of an International Public Sphere: The Reagan–Gorbachev Meetings as Televised Media Events." *Communication* 12, no. 4 (1991): 249–265.

Hays, K. Michael. "Architecture Theory, Media, and the Question of Audience." *Assemblage*, Tulane Papers: The Politics of Contemporary Architectural Discourse 27, August (1995): 41–46.

Hays, K. Michael. *Architecture's Desire: Reading the Late Avant–Garde*. Writing Architecture Series. Cambridge, MA.: MIT Press, 2010.

Hays, K. Michael. *The Event*. Studies in Continental Thought. Translated by Richard Rojcewicz. Studies in Continental Thought. Bloomington and Indianapolis: Indiana University Press, 2012.

Herscher, Andrew. "The Language of Damage." *Grey Room* 7, Spring, Special issue: On 9/11 (2002): 68–71.

Hiller, Harry H. "Toward an Urban Sociology of Mega–Events." In *Constructions of Urban Space*. Research in Urban Sociology, Volume 5, edited by Ray Hutchison, 181–205. Bingley, UK: Emerald Group Publishing, 1999.

Huyssen, Andreas. *Present Pasts: Media, Politics, Amnesia*. Stanford, Calif.: Stanford University Press, 2003.

Ibelings, Hans and Powerhouse Company. *Shifts: Architecture after the 20th Century.* Amsterdam: Architectural Observer, 2012.

Jacobs, Andrew. "China TV Network Apologizes for Fire." *The New York Times.* 10 February 2009. Accessed November 7, 2017. http://www.nytimes.com/2009/02/11/world/asia/11beijing.html?_r=0.

Johnson, Craig. "Icon and Ideology." In *Architecture, Disciplinarity, and the Arts*, edited by Andrew Leach and John Macarthur, 167–178. *Ghent*: A&S Books, 2009.

Julier, Guy. "From Visual Culture to Design Culture." *Design Issues* 22, no. 1 (2006): 64–76.

Katz, Elihu and Tamar Liebes. "'No More Peace!': How Disaster, Terror and War Have Upstaged Media Events." *International Journal of Communication* 1 (2007): 157–166.

Katz, Elihu, John Durham Peters, Tamar Liebes, and Avril Orloff, eds. *Canonic Texts in Media Research: Are There Any? Should There Be? How About These?* Cambridge, UK, Malden, MA, USA: Polity Press; Blackwell Publishers, 2003.

Kellner, Douglas. "9/11, Spectacles of Terror, and Media Manipulation: A Critique of Jihadist and Bush Media Politics." *Logos* 2, no. 1 (2003): 86–102.

Kellner, Douglas. *Media Spectacle.* London; New York: Routledge, 2003.

Kellner, Douglas. "Media Spectacle and Media Events: Some Critical Reflections." In *Media Events in a Global Age*, edited by Nick Couldry, Andreas Hepp, and Friedrich Krotz, 76–92. London and New York: Routledge, 2010.

Koolhaas, Rem. *Delirious New York: A Retroactive Manifesto for Manhattan.* New York: Monacelli Press, 1994.

Koolhaas, Rem. "White Briefs Against Filth. The Waning Power of New York." In Rem Koolhaas and Brendan McGetrick, *Content: Triumph of Realization*, 236–239. Köln: Taschen, 2004.

Koolhaas, Rem. "Kill the Skyscraper: Beijing CBD Core." In Rem Koolhaas and Brendan McGetrick, *Content: Triumph of Realization*, 473. Köln: Taschen, 2004.

Koolhaas, Rem, and Peter Eisenman. "Urgency 2007: Rem Koolhaas and Peter Eisenman. Lecture" at *cca.* June 8, 2007. Accessed November 7, 2017. http://www.cca.qc.ca/en/events/2681/urgency-2007-rem-koolhaas-and-peter-eisenman.

Koolhaas, Rem, and Brendan McGetrick. *Content: Triumph of Realization.* Köln: Taschen, 2004.

Koolhaas, Rem et al. *Considering Rem Koolhaas and the Office for Metropolitan Architecture: What Is Oma.* Edited by Véronique *Patteeuw.* Rotterdam: NAiPublishers, 2003.

Krotz, Friedrich. "Creating a National Holiday: Media Events, Symbolic Capital and Symbolic Power." In *Media Events in a Global Age*, edited by Nick Couldry, Andreas Hepp, and Friedrich Krotz, 95–108. London and New York: Routledge, 2010.

Lacayo, Richard. "The Architecture of Autocracy." *Foreign Policy* 166, May–June (2008): 52–57.

Lane, Richard J. *Jean Baudrillard*. Series Routledge Critical Thinkers. London; New York: Routledge, 2000.

Lang, Kurt and Gladys Engel Lang. "The Unique Perspective of Television and Its Effect: A Pilot Study." *American Sociological Review* 18, no. 1 (1953): 3–12.

Leach, Andrew and John Macarthur, eds. *Architecture, Disciplinarity, and the Arts*. Ghent: A&S Books, 2009.

Liebes, Tamar, James Curran, and Elihu Katz, eds. *Media, Ritual, and Identity. Communication and Society*. London; New York: Routledge, 1998.

"List of World Trade Center Tenants." CNN. Accessed November 7, 2017, http://edition.cnn.com/SPECIALS/2001/trade.center/tenants1.html.

Martin, Reinhold. "One or More." *Grey Room* 7, Spring, Special issue: On 9/11 (2002): 114–123.

Martin, Reinhold. *Utopia's Ghost: Architecture and Postmodernism, Again*. Minneapolis: University of Minnesota Press, 2010.

McLuhan, Marshall. *Understanding Media: The Extensions of Man*. New York: McGraw–Hill, 1964.

McLuhan, Marshall and Bruce R. Powers. *The Global Village: Transformations in World Life and Media in the 21st Century*. New York: Oxford University Press, 1989.

Merrin, William. *Baudrillard and the Media: A Critical Introduction*. Cambridge, UK; Malden, MA: Polity, 2005.

Ming, Xia. "Guanyu Oulu Fengge De Sikao" [Thoughts on European Continental Style]. *The Architectural Journal* 11 (1999): 44–45.

Mirzoeff, Nicholas. *The Visual Culture Reader*. 2nd edition. London: Routledge, 2002.

Mirzoeff, Nicholas. *An Introduction to Visual Culture*. London; New York: Routledge, 2009.

Mirzoeff, Nicholas. *The Visual Culture Reader*. 3rd edition. London; New York: Routledge, 2013.

Moore, Rowan. "OMA/Progress—Review." *The Guardian*. October 9, 2011. Accessed November 7, 2017. https://www.theguardian.com/artanddesign/2011/oct/09/rem-koolhaas-oma-progress-barbican.

Mosley, Jonathan and Rachel Sara. "Architecture and Trangression. An Interview with Bernard Tschumi." *Architectural Design* 83, no. 6 (2013): 32–37.

"Most Afghans Don't Even Know About 9/11, According to Disturbing Poll." *Daily Mail*. September 9, 2011. Accessed November 7, 2017. http://www.dailymail.co.uk/news/article-2035160/Most-Afghans-know-9-11-according-disturbing-poll.html.

Muschamp, Herbert. "Balancing Reason and Emotion in Twin Towers Void." *The New York Times*. February 6, 2003. Accessed November 7, 2017. http://www.nytimes.com/2003/02/06/arts/balancing-reason-and-emotion-in-twin-towers-void.html.

Neves, Joshua. "Beijing En Abyme: Outside Television in the Olympic Era." *Social Text* 107 29, no. 2 (2011): 21–46.

Nossek, Hillel. "'News Media'—Media Events: Terrorist Acts as Media Events." *Communications: The European Journal of Communication Research* 33, no. 3 (2008): 313–330.

Nossek, Hillel, Annabelle Sreberny, and Prasun Sonwalkar. *Media and Political Violence*. Cresskill, NJ: Hampton Press, 2007.

"OMA/Progress." OMA. Accessed November 7, 2017. http://oma.eu/projects/oma-progress.

One WTC. Accessed November 7, 2017. http://onewtc.com.

"One World Trade Center." *SOM*. Accessed November 7, 2017. http://www.som.com/projects/one_world_trade_center.

"One World Trade Center." *WTC*. Accessed November 7, 2017. https://www.wtc.com/about/buildings.

"One World Trade Center." *The Skyscraper Center*. Accessed November 7, 2017. http://skyscrapercenter.com/new-york-city/one-world-trade-center/98/.

Ota, Kayoko. "Toyo Ito: Big Time Dilemmas." In Rem Koolhaas and Brendan McGetrick, *Content: Triumph of Realization*, 448–449. Köln: Taschen, 2004.

Pantti, Mervi, Karin Wahl–Jorgensen, and Simon Cottle. *Disasters and the Media. Global Crises and the Media*. New York: Peter Lang, 2012.

"Plaza Details." *9/11 Memorial*. Accessed November 7, 2017. http://www.911memorial.org/plaza-details-0.

Price, Monroe E. and Daniel Dayan, eds. *Owning the Olympics: Narratives of the New China*. Series The New Media World. Ann Arbor: University of Michigan Press: University of Michigan Library, 2008.

"Projects." *Bernard Tschumi Architects*. Accessed November 7, 2017. http://www.tschumi.com/projects/19/#.

Rattenbury, Kester, ed. *This Is Not Architecture: Media Constructions*. London; New York: Routledge, 2002.

Ren, Xuefei. "Architecture as Branding: Mega Project Developments in Beijing." *Built Environment* 34, no. 4, Special issue: The Transition of Chinese Cities (2008): 517–531.

Reporters, Writers, and Editors of Der Spiegel Magazine. *Inside 9–11: What Really Happened*. New York: St. Martin's Press, 2002.

Roach, Kent. *The 9/11 Effect: Comparative Counter–Terrorism*. Cambridge: Cambridge University Press, 2011.

Roberts, John Michael, and Nick Crossley, eds. *After Habermas: New Perspectives on the Public Sphere*. Oxford: Blackwell, 2004.

Rothenbuhler, Eric W. and Mihai Coman, eds. *Media Anthropology*. Thousand Oaks, Calif.: Sage, 2005.

Rotor. Accessed November 7, 2017. http://www.rotordb.org.

Said, Edward W. *The World, the Text, and the Critic*. Cambridge: Harvard University Press, 1983.

Salomon, David L. "Divided Responsibilities: Minoru Yamasaki, Architectural Authorship, and the World Trade Center." *Grey Room* 7, spring, Special issue: On 9/11 (2002): 86–95.

Scheeren, Ole. "Culture of Change." *Büro Ole Scheeren*. July 2007. Accessed November 7, 2017. http://www.buro-os.com/culture-of-change.

Scheeren, Ole. "Made in China." *Perspecta* 39, Special issue: Re_Urbanism: Transforming Capitals, edited by Tina Di Carlo et al. (2007).

Schnoor, Christoph. "Foreword." In *Proceedings of the Society of Architectural Historians, Australia and New Zealand: 31, Translation*, edited by Christoph Schnoor, 5. Auckland, New Zealand SAHANZ and Unitec ePress; and Gold Coast, Queensland: SAHANZ, 2014.

Scott, Felicity, D. "Involuntary Prisoners of Architecture." *October* 106, Fall (2003): 75–101.

Scott, Felicity D. *Architecture or Techno-Utopia: Politics after Modernism*. Cambridge, MA: MIT Press, 2007.

Sheller, Mimi and John Urry. "Mobile Transformations of 'Public' and 'Private' Life." *Theory, Culture &Society* 20, no. 3 (2003): 107–25.

Sigler, Jennifer and Roemer Van Toorn. *Hunch 6/7: 109 Provisional Attempts to Address Six Simple and Hard Questions About What Architects Do Today and Where Their Profession Might Go Tomorrow*. Rotterdam: Episode Publishers, 2003.

Skinner, Peter and Mike Wallace. *World Trade Center*. New York, N.Y.: Metro Books, 2002.

Skinner, Peter, Mike Wallace, Valeria Manferto, and Patrizia Balocco Lovisetti. *World Trade Center: The Giants Who Defied the Skies*. Vercelli, Italy: White Star, 2002.

Sklair, Leslie. "The Transnational Capitalist Class and Contemporary Architecture in Globalizing Cities." *International Journal of Urban and Regional Research* 29, no. 3 (2005): 485–500.

Smith, Richard G. *The Baudrillard Dictionary*. Edinburgh: Edinburgh University Press, 2010.

Smith, Terry. *The Architecture of Aftermath*. Chicago: University of Chicago Press, 2006.

Sorkin, Michael and Sharon Zukin eds. *After the World Trade Center: Rethinking New York City*. New York: Routledge, 2002.

Steger, Manfred B. *The Rise of the Global Imaginary: Political Ideologies from the French Revolution to the Global War on Terror*. Oxford; New York: Oxford University Press, 2008.

Stepinska, Agnizeszka. "9/11 and the Transformation of Globalized Media Events." In *Media Events in a Global* Age, edited by Nick Couldry, Andreas Hepp, and Friedrich Krotz, 203–216. London; New York: Routledge, 2010.

Sturken, Marita. "The Aesthetics of Absence: Rebuilding Ground Zero." *American Ethnologist* 31, no. 3 (2004): 311–325.

Sudjic, Deyan. *The Edifice Complex: How the Rich and Powerful Shape the World*. London; New York: Allen Lane, 2005.

Tafuri, Manfredo. *Architecture and Utopia: Design and Capitalist Development*. Cambridge: MIT Press, 1976.

Tafuri, Manfredo and Francesco Dal Co. *Modern Architecture*. 2 volume History of World Architecture. New York: Electa/Rizzoli, 1986.

Taylor, Charles. *Modern Social Imaginaries*. Public Planet Books. Durham: Duke University Press, 2004.

Taylor, Charles. *A Secular Age*. Cambridge, MA: Belknap Press of Harvard University Press, 2007.

Trachtenberg, Marvin and Isabelle Hyman. *Architecture, from Prehistory to Post-Modernism: The Western Tradition*. New York: H.N. Abrams, 1986.

Tschumi, Bernard. *Architecture and Disjunction*. Cambridge, MA: MIT Press, 1994.

Tschumi, Bernard. *Questions of Space: Lectures on Architecture*. London: Bernard Tschumi and the Architectural Association, 1995.

"Bernard Tschumi. Advertisements for Architecture." *WFW*, December 7, 2010. Accessed November 7, 2017. http://www.we-find-wildness.com/2010/12/bernard-tschumi-advertisements-for-architecture/.

Tschumi, Bernard and Max Protetch Gallery. *The Manhattan Transcripts*. London and New York: Academy Editions, St. Martin's Press, 1981.

Tschumi, Bernard and Museum of Modern Art (New York). *The Manhattan Transcripts*. London; New York, NY: Academy Editions; Distributed to the trade in the USA by St. Martin's Press, 1994.

Tschumi, Bernard and Nederlands Architectuurinstituut. *Bernard Tschumi: Architecture in/of Motion*. Rotterdam: NAi Publishers, 1997.

Tschumi, Bernard and Irene Cheng. *The State of Architecture at the Beginning of the 21st Century*. Series Columbia Books of Architecture. New York: Monacelli Press, 2003.

Tuck Wong, Shue and Sun Sheng Han. "Whither China's Market Economy? The Case of Lijin Zhen." *Geographical Review* 88, no. 1 (1998): 29–46.

Vidler, Anthony. *Architecture between Spectacle and Use*. Williamstown, MA. New Haven: Sterling and Francine Clark Art Institute; Distributed by Yale University Press, 2008.

Vriesendorp, Madelon, Shumon Basar, Stephan Trüby, and Architectural Association (Great Britain). *The World of Madelon Vriesendorp: Paintings, Postcards, Objects, Games*. London: AA Publications, 2008.

Watkin, David. *A History of Western Architecture*. London: Barrie & Jenkins, 1986.

Weimann, Gabriel and Conrad Winn. *The Theatre of Terror: Mass Media and International Terrorism*. New York: Longman, 1994.

Wergeland, Even Smith. "When Icons Crumble–the Troubled Legacy of Olympic Design." *Journal of Design History* 25, no. 3, Special issue: Design Histories of the Olympic Games (2012): 304–318.

Wigley, Mark. "The Translation of Architecture, the Production of Babel." *Assemblage* 8, February (1989): 6–21.

Wigley, Mark. "Insecurity by Design." In *After the World Trade Center: Rethinking New York City*, edited by Michael Sorkin and Sharon Zukin, 69–85. New York: Routledge, 2002.

"World Trade Center Timeline of History." *World Trade Center*. Accessed November 7, 2017. http://www.wtc.com/about/history.

Xinhua. "Beijing's New Ring Road to Run 1,000km." *China.org*. July 7, 2014. Accessed November 7, 2017. http://www.china.org.cn/china/2014-07/07/content_32881371.htm.

Zelizer, Barbie and Stuart Allan. *Journalism after September 11*. 2nd edition. London; New York: Routledge, 2011.

Zhang, Jingxiang and Fulong Wu. "Mega–Event Marketing and Urban Growth Coalitions: A Case Study of Nanjing Olympic New Town." *The Town Planning Review* 79, no. 2/3, Special issue: Planning the Chinese City (2008): 209–226.

Zhu, Tao. "Building Big, with No Regret." *Architectural Association School of Architecture* 63 (2011): 104–110.

Zhu, Ying. *Two Billion Eyes: The Story of China Central Television*. New York, Jackson, TN: New Press; Distributed by Perseus Distribution, 2012.

Žižek, Slavoj. *Welcome to the Desert of the Real*. New York, NY: Wooster Press, 2001.

Žižek, Slavoj. "Badiou 1: Event and Atc." *In Philosophy: Spinoza, Kant, Hegel and ... Badiou!* 2007. Accessed November 7, 2017. http://www.lacan.com/zizphilosophy3.htm.

Žižek, Slavoj. *Event: A Philosophical Journey through a Concept*. Brooklyn, London: Melville House, 2014.

Zukin, Sharon. "Our World Trade Center." In *After the World Trade Center: Rethinking New York City*, edited by Michael Sorkin and Sharon Zukin, 13–22. New York: Routledge, 2002.

PART 2

Architectural Imaginings and Colonial and Post-colonial Realities

CHAPTER 4

Modernism as a Mechanism of Power and Control in Colonial Contexts: The Project of Modernity in Cape Town, South Africa

André van Graan

The Discourse of Colonial Modernism

The discourse of modernism in architecture, particularly in a colonial context, is strongly grounded in theories of power, power relations, mechanisms of control, and terrains of contestation and negotiation of difference. Modernism was a discourse grounded in a paradigmatic shift in thinking about the way that space was constructed and perceived and established a new modernist architectural discourse. The shift reunited a schism that had developed between spatial production related to function and building form or style.

Twentieth century theorists such as Michel Foucault and Henri Lefebvre have re-asserted space into the discourse of social science and history, posing alternative conceptions of space that highlight the significance of space in social discourse and reformulate the nature of space itself. Jordaan in her examination of the agency of the architect in post-apartheid South Africa refers to tangible and finite space. She alludes to the manner in which places are often designed in denial of reality.[1] The examination of the spatial dialectic is a discourse; every discourse operates through language. Michel Foucault developed the concept of a 'discourse' in *The Archaeology of Knowledge*. For Foucault, everything exists within a discourse: it is contingent on it and part of it.[2] Where he refers to 'statements' within a discourse,[3] these could be books, articles, clients, or buildings: a discursive formation brings these diverse elements together, but can never be definitive, as it is always provisional. Therefore, Foucault's

1 June Jordaan, "Architectural Agency and 'Place-Making' in a Transformative Post-Apartheid South African Landscape," in this volume.

2 'Discourse [is] a group of statements in so far as they belong to the same discursive formation; it does not form a rhetorical or formal unity, endlessly repeatable, whose appearance or use in history might be indicated...; it is made up of a limited number of statements for which a group of conditions of existence can be defined.' Michel Foucault, *The Archaeology of Knowledge*, trans. Alan M. Sheridan (London: Routledge 2004), 131.

3 Ibid., 30–31.

 | DOI 10.1163/9789004376793_005

archaeology is a useful means to evaluate the products of culture, as it does not imply any definitive or specific genesis of these products.

Foucault also makes the link between the understanding of discourse as expressed through architectural culture and the significance of the production of space. He focuses attention on the spatiality of the actually lived and socially produced space of sites and the relations between them and refers to these real, heterogeneous spaces as 'heterotopias' in contrast to planned 'utopias'—the places designed in denial of reality as Jordaan suggests.[4] Foucault regarded architecture as 'not only an element in space, but ... [it] is especially thought of as a plunge into a field of social relations in which it brings certain effects.'[5] This is fundamental to an understanding of architecture as the visible politics of a colonial society. In Europe, modernism was the vehicle for developing a new socialist architectural vision of a utopian egalitarian society, based on broad Marxist principles. This restructuring of society was manifested in social programmes involving urban housing, healthcare, and education schemes, in addition to creating new workplaces for a rising urban proletariat.[6] There were often contradictory forces at play—the incorporation of socialist ideals while simultaneously seeking to avoid the schismatic disruption of a socialist revolution. Le Corbusier suggested in his *The Radiant City* in 1933:

> From the huddle of hovels, from the depths of grimy lairs ... there sometimes came the hot gust of rebellion; the plot would be hatched in the dark recesses of an accumulated chaos in which any kind of police activity was extremely difficult.... St. Paul of Tarsus was impossible to arrest while he was in the slums ...[7]

With these words, Le Corbusier reflected the fear that existed among the middle classes of the threat of revolt that the slums offered. His idea to 'clear the decks completely and replace the centre of Haussmann's city with one built with control and hierarchy in mind'[8] was the universal Zeitgeist in the 1930s.

4 Jordaan, "Architectural Agency and 'Place-Making'."

5 Michel Foucault, *Power*, ed. James D. Faubion, trans. Robert Hurley and others (London: Penguin Books 1994), 362.

6 Sibel Bozdoğan, *Modernism and Nation Building. Turkish Architectural Culture in the Early Republic* (Seattle: University of Washington Press, 2001); Elizabeth Darling, *Re-forming Britain. Narratives of Modernity before Reconstruction* (London: Routledge, 2007).

7 Cited in James C. Scott, *Seeing like a State. How Certain Schemes to Improve the Human Condition Have Failed* (New Haven: Yale University Press, 1998), 116.

8 Id., 116.

One finds programmes being undertaken around the world to implement slum clearance and new housing programmes that were, if not in their initial proposals, more often than not, linked to the removal, segregation, and control of the urban poor. Although social reform is the often cited reason, one wonders whether the aftermath of the First World War and the Bolshevik revolution in Russia was not feared as a threat to established order and whether they led to new Foucaultian methods of social control and surveillance to avert any possible threat. Modernism became the mechanism for visibly demonstrating change: the visualisation of political will.

The project of modernity was manifested in Cape Town, South Africa, through reliance upon scientific, technological solutions to spatial and social concerns. The dominant instrumentalization of segregation, whilst initially a racist values-based project, was both bureaucratic and technocratic. Therefore, its spatial manifestation was also a technocratic one: rather than the ambiguous, negotiated manifestations of the nineteenth century, those of the twentieth century were precise and produced initially through the work of the planners, rather than the inhabitants, of space.

During this period, there were also significant changes in the hegemonic perceptions of the city, and this was substantiated by legislation introduced by the local city council. For example, the eradication of slums in Cape Town in the early 1930s served as a catalyst for spatial change and the dislocation of lived space as the city sought to re-create itself as a modern, rationally planned metropolis. Architects sometimes appeared to use the criteria of the modernist discourse as a mechanism to wield social control on marginalized members of society creating mechanisms of removal, exclusion, surveillance, and control based on ethnicity.

Modernism and Power Relations: Heterogeneity and Identity

It is important to delve into colonialism's history 'not only to document its record of domination but also to track the failures, silences, displacements and transformations produced by its functioning.'[9] More than a unidirectional and successful imposition, the 'export of notions, systems, and practices which displace[d] indigenous forms or recreate[d] them in the image of the colonial power' faced negotiation, contestation, and re-constitution in colonial

9 Garth A. Myers, "Intellectual of Empire: Eric Dutton and Hegemony in British Africa," *Annals of the Association of American Geographers* 88, no. 1 (1998): 1–27.

societies.[10] This negotiation and contestation is enough in evidence to say that the spatial project cannot be reduced to mere 'self-expression'.[11]

In his paper *Of Other Spaces*, Foucault described the twentieth century as an era of juxtaposition of differences. The present time he considered to be an age of space. Space, in his terms, was heterogeneous, and he speaks of living 'inside a set of relations that delineates sites which are irreducible to one another and absolutely not superimposable on one another.'[12] These differing conditions of existence he defines as *utopias* and *heterotopias*. Heterotopias in turn are divided into heterotopias of crisis and heterotopias of deviation. This concept of heterotopias is a useful mechanism for defining the way in which modernism as a societal and aesthetic construct was negotiated and contested in colonial and post-colonial architecture. Again, Jordaan speaks of characteristics of place-making that are infinite, open, and fluid, in contrast to finite, bounded, and differentiated spaces.[13] This latter characteristic of controlled finite places with an identifiable character finds resonance in the modernist housing projects of Cape Town.

Underlying studies of colonial structures is the study of power. Foucault demonstrates that it is not that power becomes a production of design professionals, but rather space a realm appropriated by politics. Social operations, productions of power, are conducted *through* and *in* the realm of space. He points out: 'After all, the architect has no power over me. So he should be placed in another category—which is to say that he is not totally foreign to the organisation, the implementation, and all the techniques of power that are exercised in a society.'[14] According to Foucault, space is itself not exclusionary or prohibitionary:

> Nothing is fundamental. That is what is interesting in the analysis of society.... These are not fundamental phenomena. There are only reciprocal relations, and the perceptual gaps between intentions in relation to one another.[15]

Modernism in the colonial context led to the creation of hybrid identities that were based on power relations. In South Africa, with the strengthening

10 Id., 7.

11 Ibid.

12 Michel Foucault, "Of Other Spaces," *Diacritics* 16, no. 1 (1986): 23.

13 Jordaan, "Architectural Agency and 'Place-Making'."

14 Foucault, *Power*, 357.

15 Id., 356.

of Afrikaner Nationalism and increasingly overt racism in the period under consideration, a simple binary division of Self and Other is inadequate as a basis for the multiplicity of identities that emerged or were officially created. The notion of a heterogeneous, unequal society was to be reflected in the architectural use of modernism. Post-colonial writing often debates the impact of modernism on colonial societies and writers, such as Sangari, consider it to be a significant act of cultural self-definition within these societies.[16]

The dominant discourse of British imperialism waned in the inter-war period in South Africa, creating the space for the expression of different identities through architecture. Within the debate around eclecticism and heterotopic spaces, hybrid identities emerge as significant factors that need to be considered. The South African academic, Desiree Lewis, says in her article 'Writing Hybrid Selves' that both the terms hybridity and hybridization are used to describe the production of differing identities and questions the idea of polarized, completely essentialized identities.[17] Spivak questions the validity of simple binary sub-division of

> ... the remotely orchestrated, far-flung, and heterogeneous project to constitute the colonial subject as Other. This project is also the asymmetrical obliteration of the trace of that Other in its precarious Subjectivity. It is well known that Foucault locates epistemic violence, a complete overhaul of the episteme, in the redefinition of sanity at the end of the European eighteenth century. But what if the two projects of epistemic overhaul worked as dislocated and unacknowledged parts of a vast two-handled engine? Perhaps it is no more than to ask that the subtext of the palimpsestic narrative of imperialism to be recognised as 'subjugated knowledge'.[18]

Although Spivak refers specifically to colonial discourse here, it is equally possible to apply it to the modernist episteme that denied the validity of any subtext that was disqualified by the tenets of mainstream modernism.

In considering the application of these theoretical concepts, James C. Scott's critique of what he calls 'The High Modernist project' is a significant channel that connects more abstract theory to its application within a discourse.

16 Kumkum Sangari, "The Politics of the Possible," *Cultural Critique* 7, Fall (1987): 157–186.

17 Desiree Lewis, "Writing Hybrid Selves: Richard Rive and Zoë Wicomb," in *Coloured by History, Shaped by Place*, ed. Zimitri Erasmus (Cape Town: Kwela Books, 2001).

18 Gayatri C. Spivak, "Can the Subaltern speak?" in *The Post-Colonial Studies Reader*, eds. Bill Ashcroft, Gareth Griffiths, and Helen Tiffin (London: Routledge, 1995), 24–25.

He explains that this project combines three elements: the administrative ordering or rational engineering of nature and society; the unrestrained use of power to achieve these designs; and a weak or emasculated civil society that is unable to resist these plans. He says: 'The ideology of the high modernism provides, as it were, the desire; the modern state provides the means of acting on that desire; and the incapacitated society provides the levelled terrain on which to build (dis)utopias.'[19]

The indigenous population in South Africa formed a completely subordinate class against whom all manner of prejudices and preconceptions were held. At the Cape, the so-called 'Coloureds' were seen as an intermediate grouping that held a position above that of the Africans—but below that of the Whites. These relationships always had a spatial fix in terms of people's residential spaces in the city, although there was a great deal more mixing, particularly among the poorer classes, and this was generally tolerated in the nineteenth century.

This emphasis of the supposed superiority of the European settlers led to the justification of many decisions that were made regarding the role, living conditions, and spatial segregation of those deemed 'Other'. The fear of mixing was a constant theme in the local media, and the concern was expressed about the potential outcome of Africans being allowed to live in the city:

> A striking proof is given by certain statistics that have been published by the Association for the Prevention of Consumption: these seem to show conclusively that as the Kafir dons the white man's clothing and lives under conditions proper to the European, he becomes increasingly susceptible to the white man's scourge, which is decimating his race and tends to discount the large increase of population that takes place within the native territories. It is manifest that the evils, which thus beget our native populations transferred from reserves to the towns, are bound to have a disastrous effect upon the white population also.[20]

This extremely racist point of view would seem to link in with the outcome of the Stallard Commission and lend credence to its outcomes. The Commission had, in 1923, proposed that urban areas were to be reserved as 'White' areas, and that Africans could come to the cities to work, but that they were to be treated as temporary residents and not allowed to bring their families with them. Architecture served the political ends of government. This attitude to

19 Scott, *Seeing like a State*, 89.

20 "The Urban Problem," *The Architect, Builder & Engineer*, April 1923, 17.

people who were not white, informed decisions on slum clearance and housing, along with preferential labour policies and the exclusion of people from decisions regarding their living and working conditions.

In order to locate modernism in Cape Town, it is important to consider the relationship between the city and the state, and the state as a former British colony and Britain, as this sets the framework within which the modernist discourse was located. As a late-colonial regime with a uniquely racist constitution at the time, the significance of ethnicity and the role it played in the modernist project is very important. The concept of the colonial 'Other' forms part of post-colonial theory and explains the manner in which those people (the colonised) were dealt with as outsiders. In South Africa, in the early twentieth century, the simple division of people into an 'Us', who were White, and an 'Other', who were not, multiplied into a much more heterogeneous cluster of sub-divisions, taking the racist discourse to new, intricately nuanced levels. Afrikaner nationalism developed considerably in this period as the Afrikaners took ever-stronger positions in the government of the country until, in 1948, they achieved outright control. Their desire to alienate themselves from anything British led to them taking on modernism at an early stage as it represented a more neutral basis for self-expression, much as Zionism was to do in Israel.[21] Political modernism was adopted at an early stage in the Union of South Africa and with it came the concept of racial segregation through the expedient mechanism of slum elimination and town planning.

The Cape Town City Council used segregation as a method of controlling the living spaces of those whom they wished to control to remove them to places where they would be clearly visible. Foucault suggests that 'all authorities exercising individual control function according to a double mode; that of binary division and branding ... and that of coercive assignment, of differential distribution (who is he; where must he be; how is he to be characterised; how is he to be recognised...).'[22] The colonial discourse on racism focussed heavily on differentiated assignment particularly with regard to the 'superior' characteristics and 'responsibility' of Whites in relationship to other ethnic groups.

21 Alona Nitzan-Shiftan, "Contested Zionism-Alternative Modernism: Erich Mendelsohn and the Tel Aviv Chug in Mandate Palestine," *Architectural History* 39(1996): 147–180. 'The modern movement's rupture from the past was amplified in the Zionist context ... modernism signified a break with the anti-Semitism of their subordinated life in exile. It symbolized the transition from centuries of being a stereotyped minority in Europe to the promise of constructing an autonomous Self.'

22 Michel Foucault, *Discipline & Punish: The Birth of the Prison*, trans. Alan Sheridan (New York: Vintage Books, 1995), 199.

This can be linked to the misinterpretation of the Nietzschean concept of the *Ubermensch*, as happened in Nazi Germany, by linking superiority with a particular ethnic group. The 'White Man's Burden' was the description used in the local journal *Architect, Builder & Engineer* in 1927 to predicate a diatribe on other ethnic groups:

> We find for instance that the native labour is unintelligent and unskilled, that the Coloured labour is only partly skilled, that the native labour though loyal is incompetent, and the Coloured worker, though partially competent, is untruthful and disloyal. It is our duty and our privilege to render such service to the measure of our own capacity ourselves. That is the White man's burden. It is the White man's burden to so minister to the weakness of subordinate races and so to utilise it even when to his own advantage, that no cruel injustices shall result.[23]

Shamil Jeppie identifies 'a powerful racial element in the discourse on housing and urban renewal.'[24] According to him, '"Slums" and "Coloured" were almost interchangeable terms.'[25] Richard Whiteing describes the intermediate or residual position held in the Cape by Coloured people, whose position is 'assigned' on the basis of them being neither White nor African. This identity had negative and derogatory connotations: 'At the core of these connotations was the idea that the Coloured people were the result of miscegenation and were racially impure.'[26] This reiterates Foucault's description of the identification and exclusion of the leper.[27] The *Architect, Builder & Engineer* complained:

> What are these centres doing for their non-European inhabitants, citizens, (and) ratepayers? Very little we are afraid. The local coloured press is getting quite heated up about it, and there is considerable justification… They are very law-abiding as a whole. They serve us well. We ought for their sakes and our own to see that they get what is their due. There is a sad lack in our centre of proper housing for them, decent amenities

23 "The White Man's Burden," *The Architect, Builder & Engineer*. June 1927, 2.

24 Shamil Jeppie, "Modern Housing in the District: The Canterbury and Bloemhof Flats," in *Recalling Community in Cape* Town, eds. Ciraj Rassool and Sandra Prosalendis (Cape Town: District Six Museum, 2001), 117.

25 Ibid.

26 Richard Whiteing, "The Emergence & Development of a Distinct Coloured Identity," MPhil. Diss., University of Cape Town, South Africa, 2003.

27 Foucault, *Discipline & Punish*, 198.

> by way of playing fields, bathing pavilions, and the like... The sooner we realise that these folk do not wish to flout our prejudices[28]

The somewhat patronising tone of the article clearly identifies 'Coloured' people as 'Other'. Their perceived role in society is also clearly stated: 'They serve us well.' The 'Coloured' category was further sub-divided in the prevailing colonial ethnic discourse. The 'Malay' section of the community was viewed quite differently to the 'Cape Coloured'. In the published assessment of the Schotsche Kloof housing design competition for Cape Town's Bo Kaap,[29] they are described as: 'As a race or caste or class or faith, call it what we may ... noted as being intelligent, sober, law-abiding, cleanly, and in the vast majority of cases anything but unskilled.'[30]

The outcome of this differentiation and assignment of 'Otherness' to ethnic groups was used to map out the differing planning approaches that would be adopted by the City Council in the housing policies applied to different groups.

Modernism as a Mechanism of Social Engineering

The Cape Town City Council had used the Public Health Act of 1897 to move Africans from District Six in the city to a colonial government camp at Uitvlugt forest station in 1901, which forming the basis for the settlement known as N'dabeni. This was established in terms of the Locations Act of 1901, which is significant in that it was the first act in South Africa to legislate for racially segregated so-called 'townships'.[31] The Slums Act No. 53 of 1934 called for the identification of slum buildings and areas and the notification of owners to repair or demolish the identified properties or face expropriation.

In attempting to deal with slum clearance, the city of Cape Town embarked on an ambitious housing programme that was largely focussed on the Coloured population. They created a series of housing projects that focused on

28 "Our Slum Problem," *The Architect, Builder & Engineer*, June 1936, 16.

29 The Bo-Kaap was also known in the colonial period as the 'Malay Quarter' and lies to the west of the city centre against Signal Hill. In the 19th century it was largely (but not exclusively) settled by slaves freed in terms of the Abolition of Slavery Act of 1834. A large section of the population were (and still are) largely Muslim.

30 "Schotsche Kloof Housing Competition," *The Architect, Builder & Engineer*, April 1939, 2.

31 Townships is the name applied in South Africa to planned settlements, usually for African people, but the term was also applied to areas designated for any so-called 'non-white' people.

the provision of flats, both in cleared areas of the inner city and within new residential areas on the Cape Flats.

Housing legislation was used in Cape Town to implement the City Council's policy with regard to the identification, classification, and rehousing of the city's poor. Scott and Soja both draw attention to the modernist programme of dealing with the poor of the world in a much more comprehensive way than had ever been undertaken before.[32] Poverty, ethnic difference, and slums represented a threat to a controlled colonial society. The legislation rendered visible the problems of the city's slum areas and, furthermore, created environments where the poor were either clearly visible, for example in the inner city schemes with their open spaces, easy access, and, hence, controllable spaces, or removed to separate areas, isolated from the White populace and, thus, like Foucault's leper, identified and excluded. The first level of the process of control was the dehumanisation of the problem. Thus, in the late 1920s, there is a call for 'comprehensive housing schemes for various Classes'.[33] No mention was made of who was being classified.

It was often perceived that the slum problem in South Africa differed from that in 'European countries'. The Vice Chairman of the Rehousing and Slum Elimination Association, F. Gebhardt, explained in 1936 that: 'The psychological factors that tend to create slums ... apart from extensive unemployment ... and the meagre earnings of the lapsed masses ... is the powerful attraction which "White" civilisation exerts upon the more primitive races.'[34] He goes on to reiterate the White supremacy ideology as he points out that 'While slums, therefore, constitute a problem of the poorest classes, irrespective of colour, it is incumbent on the White races to solve the problem.'[35] But there were contradictory views at the time as to whether the problems of slums were different in South Africa or whether the problem was universal. Rather oddly, in view of his later role in defining segregation based on difference, Dr. D.F. Malan, the subsequent National Party leader and prime minister after 1948, said at a housing conference held in Cape Town in 1930 that 'the crux of the problem in connection with housing is the provision of housing for the very poorest classes', and pondered what to do with the existing slums and the overcrowding that

32 Scott, *Seeing like a State*; Edward Soja, *Post-modern Geographies. The Reassertion of Space in Critical Social Theory* (London: Verso, 1989).

33 Cape Town, Housing & Estates Committee minutes, August 30, 1927.

34 F. Gebhardt, "Housing Issues," *The Architect, Builder & Engineer*, January 1936, 13.

35 Ibid.

existed. In getting rid of slums, he referred to legislation in the United Kingdom that should be followed as he considered conditions in South Africa to be no different to those in the UK.[36]

The evidence, thus, suggests that the Slums Act was not as effective a tool as the Council had imagined it to be in the elimination of overcrowding and slums and that there was now a perceived need to tackle a much more comprehensive planning exercise that would create a modern city unencumbered with the problems of the past. These views, calling for a more comprehensive way of dealing with slums, reiterates the views of Le Corbusier. Scott indicates the attitude of Corbusier who condemned the 'misery', 'confusion', 'rot', 'decay', 'scum', and 'refuse' that he believed needed to be overcome. He regarded the slum inhabitants as 'a dead weight on the city, an obstacle, a black clot of misery, of failure, of human garbage'.[37] He objected to the lack of discipline of these people, which he regarded as being against nature, which was all discipline. He was also concerned about the potential revolutionary threat posed by these people. Scott says that Corbusier understood, as Haussmann had, that crowded slums were and had always been an obstacle to efficient policing. He goes on to say: 'Le Corbusier proposed to clear the decks completely and replace the centre of Haussmann's city[38] with one built with control and hierarchy in mind.'[39] Foucault's description of panopticism is the underlying order of the modern city as proposed by the modernist vision of Le Corbusier. Thus, one finds great concern being expressed in Cape Town in 1935 regarding 'the distressing and alarming position created by the racial mixtures in our population'.[40] Neither the State nor local authorities, it was contended, 'fully appreciate the dangers which may be apprehended when economic pressure compels people of European descent to live in close proximity to those of other races. The peril is by no means confined to the spread of disease and crime; it may threaten White civilisation in the continent of Africa.'[41]

It is clear that the agenda of the City was to overcome both overcrowding and racial mixing—characteristic of the inner suburbs—by establishing cheaply planned settlements on the Cape Flats, well away from White residents.

36 Cape Town Housing & Estates Committee minutes, February 18, 1930.

37 Scott, *Seeing like a State*.

38 Baron Haussmann was the nineteenth century planner who radically re-planned the centre of Paris adding broad boulevards.

39 Scott, *Seeing like a State*, 116.

40 "Cape Town's Coloured Problem," *The Architect Builder & Engineer*, December 1935, 22.

41 Id., 22.

The Provincial Administrator Fourie, in examining Cape Town's housing proposals for an area adjacent to Maitland Garden Suburb did not like

> [t]he scattering of non-Europeans from one end to the other of the city [which] was not in the best interests of Cape Town.... The City should be completely zoned ... so that certain sections should be set apart as European areas, others as Non-European areas and areas where noxious trades might be established.[42]

This, *de facto*, creates the model of an ethnically segregated city that was to be reinforced in 1950 by the Group Areas Act. Scott in his description of 'Authoritarian High Modernism' observes that:

> Colonial regimes, particularly late colonial regimes, have often been sites of extensive experiments in social engineering. An ideology of 'welfare colonialism' combined with the authoritarian power inherent in colonial rule have encouraged ambitious schemes to remake native societies.[43]

Despite opposition from the African residents, the Council went ahead with its plans and in 1927 opened Langa Township. They appointed the same Town Planner who had been responsible for the layout of Pinelands, a White residential area, Longstreth Thompson, to prepare the layout along similar Garden City principles.

This new residential area was developed with a focus on the provision of single quarters, as the Council sought to emphasize the notion that Africans were 'temporary' dwellers in the city. In the design of the housing provided, some were designed by architectural practices, while others were developed by the newly formed Housing Department of the City Council. The designs varied considerably in the way that they used modernism, and it is evident that public housing formed a terrain of negotiation of modernism. This relates to what was happening in Britain, the USA, and Turkey, thus, reflecting an international tendency where the acceptance of modernism for housing was often contested. The Africans, removed as far as possible from family structures, were given mainly hostel accommodation:

> Extensive work will shortly be commenced at the new native location just outside Cape Town. Mr. J. Thompson of Garden City fame is the

42 Cape Town Housing & Estates Committee Minutes, May 10, 1928.

43 Scott, *Seeing like a State*, 97.

> architect and it is intended to erect markets, hostels, a police station and colleges. Of course, all of this work will be more or less of a simple nature, but it is likely to run into a considerable figure: two of the hostels alone have to house two thousand natives each.[44]

The essential characteristic of Langa, despite the subsequent adoption of modernist design principles, was the acceptance of the principle as laid down by the Stallard Commission[45] that Africans were only temporary urban residents and that any family life was to be discouraged—families remaining in rural areas—while men were accommodated in hostel blocks. Racism remained the underlying concern in the design of the hostels. The design of the residential blocks reflect the prevailing modernist principles that had been applied to projects in Germany, France, and Britain. (Figure 4.1) The inherent inhuman principles of segregation and separation and the negation of the family is not apparent in the slick architectural images of the mayoral minutes.

In District Six, modernism's 'clean knife' would be used to cut out the decay and to replace the existing living pattern with a 'modern' vision of urban housing, that laid open and visible the threatening alleys and spaces that potentially fomented revolution. The Canterbury-Bloemhof flat scheme in District Six was set in open spaces, clearly visible and controllable.

In the Bo-Kaap, there was a reticence to adopt the same approach despite the declaration of slum areas and the palpable air of decay. Schemes that had been prepared for replacing the existing urban fabric with a 'modern' alternative were shelved, and the Schotsche Kloof apartment scheme of 1944, the outcome of an architectural design competition, was prominently placed on a green field site above the historic area of this predominantly Muslim area. (Figure 4.2) In the design of the Schotsche Kloof Flats, the architects proposed a smaller living unit, 'a simplified flat' where the sleeping spaces were divided by a screen wall rather than a full height division. This was the cause of considerable controversy. It found favour with the Council on economic grounds, but some councillors objected on the grounds that it was unsuitable for 'Malays

44 "New Native Township," *The Architect, Builder & Engineer*, October 1923, 33.

45 In 1923, the Natives (Urban Areas) Act laid down that all Africans should be segregated in locations. Control on movement was also introduced and later became known as 'influx control'. The act was based on the principle that Africans should remain temporary urban dwellers only. In 1922, the Stallard Commission had stated that: 'The Native should only be allowed to enter the urban areas, which are essentially the White man's creation, when he is willing to enter and then minister to the needs of the White man, and should depart there from when he ceases so to minister'.

FIGURE 4.1 *'Bachelor' quarters for African men in Langa, Cape Town, South Africa. Cape Town City Council Mayoral minutes 1942*
SOURCE: CITY OF CAPE TOWN.

[who] on the whole earned considerably more than the average Coloured person.'[46] It was suggested that the 'Architect had put forward a design which was in vogue for superior Europeans.'[47] This proposal is closely linked to the *Existenzminimum* that was the subject of CIAM 2 Congress (Congres Internationale des Architecture Moderne) held in Frankfurt in 1929.[48]

Ambitious residential projects were proposed for the Cape Flats, far removed from White areas, and the planning of areas, such as Q-Town,[49] reflected modernist planning based on ideas emanating from Germany where the

46 Cape Town Housing & Slum Clearance Committee minutes, October 11, 1939.

47 Ibid.

48 Eric Mumford, *The CIAM Discourse on Urbanism 1928–1960* (Cambridge, Mass: MIT Press 2000).

49 Alicedale and Q-Town (or Kewtown) were planned settlements for people of colour on the Cape Flats—the sandy area susceptible to winter flooding that lies to the east of the city.

FIGURE 4.2 *Schotsche Kloof Flats in 2017, Cape Town, South Africa*
PHOTO: ANDRÉ VAN GRAAN.

Frankfurt housing of Ernst May extended the modernist discourse in housing to create improved living conditions for the city's population.

On the Cape Flats, the Kewtown (initially Q-Town) flats were being built with the first block ready for occupation in 1941. It had been suggested at the time that this would become 'a town the size of Bloemfontein'.[50] According to Jeppie, '[c]ity Planners believed that all that was actually needed was actually putting together a few blocks of low-cost flats.'[51] Although this is Jeppie's contention, there are in fact broad records in the City Council's Mayoral Minutes of the extensive proposals for this development; it would appear to have been an extremely ambitious scheme. By 1944, Kewtown had 224 flats with double that number out on tender. The layout ominously reflected Le Corbusier's planning layout as interpreted for central Cape Town in 1938, separated, not by parks, but by acres of concrete and tarmac. (Figure 4.3) These concepts were to inform planning decisions in Cape Town as slum clearance schemes forced people of colour out of the city. The link between the modern housing and mechanisms of control were clearly apparent. The spaces were rendered visible.

50 Jeppie, "Modern Housing in the District," 123.
51 Ibid.

FIGURE 4.3 *Implementing the Corbusian vision for a marginalized community: Kewtown, Cape Town, South Africa. Cape Town City Council Mayoral Minutes 1943*
SOURCE: CITY OF CAPE TOWN.

Conclusions

The architecture of the city housing projects mediates between programmatic requirements and site. Possibly, the economic constraints of the projects led to a 'purer' form of modernism being applied. The schemes were clearly part of a comprehensive strategy for planning the city. The strategies for the implementation of the housing schemes were extensions of social and political domination by the spatial control of people by the city. Whether the architects were consciously part of the process or merely the instruments is hard to say from the available evidence. They appear to be applying norms that had been used in more egalitarian societies, but as Bozdoğan[52] points out, architectural form does not necessarily reflect only one political ideology. It is apparent that the forms and layouts adopted make access, visibility, and observation easier. Was this intentional or inherent in the design decisions taken? They seem to reflect a *Zeitgeist* that placed society above the individual, and the State above society. Individual identity is subservient to the needs of the authorities.

52 Bozdoğan, *Modernism and Nation Building*, 10.

The last of Jeppie's assertions, namely, the use of housing as a laboratory for the promotion of modernity, certainly occurred in the Schotsche Kloof flats with the introduction of new unit types that had not been used previously. Thus, it would appear that the schemes corroborate the critique made by Jeppie. They strongly reflect the mediated modernism espoused by Bozdoğan rather than the avant-garde modern movement ideals of Le Corbusier or the German architects.

The manner in which South Africa developed constitutionally in the early twentieth century set the stage for the introduction of modernism. The balance between English-speakers and Afrikaners was always in a state of tension in the period prior to the Second World War. The opposition under General Herzog and later Dr. D.S. Malan, whose Nationalist Party swept to power in 1948, was constantly challenging the government of Louis Botha and Jan Smuts and their South African Party. All issues were negotiated in the precarious balance of power that implemented programmatic modernism within a colonial context. The colonial 'Other', a binary division between the White colonists and people of colour, considered as 'Other', also broke down into a much more heterogeneous grouping with sub-groups being created on both sides. Jewish people formed a sub-group at a time when Nazism was embarking on its programme of genocide in Europe and they responded differently to modernism compared to English-speakers who retained strong links to Britain. Afrikaner Nationalism grew considerably during this period as they developed into a powerful sub-grouping, again with their own interpretation of the modernist discourse.

The deliberate omission of people of colour from the political arena also further enhanced the colonial racial divide of the nineteenth century and embedded prejudices in legislation. As the country adopted modernist planning principles in the 1920s, this programmatic modernism created racially segregated residential areas, thus, separating people in the places where they lived and the manner in which they lived, as housing schemes were designed for people of colour removed from the older historically more integrated living areas of the nineteenth century. Both the State and the City were implicated in this process with legislation introduced by the State leading to architecture of segregation and making segregation a lived reality long before the Nationalist government introduced the Group Areas Act in 1950.

Health and planning were instruments used to negotiate difference, politically and socially, and affected the manner how space was produced as the city changed from a colonial city into a modern city by the Second World War. The adoption of the flat as the basic residential unit for Coloured people and the bachelor quarter for Africans that occurred during this period was a stark contrast to the individual house that was preferred for Whites. State and City negotiated modernism through planning and architecture on a scale that was to re-create Cape Town as a modern city—the outcome of a process of social engineering.

Public housing schemes reinforced one dimension of the topography of segregation in the city. The political ideology of apartheid could be seen to be essentially a system of spatial exclusion, based on ethnicity and an ethnic hierarchy. What is surprising is the extent to which this system and the establishment of the segregationist topography was in place prior to the Second World War, and how closely it was linked to the provision of housing by the local authorities. In the aftermath of legislated discrimination, the complex socio-political and ideological framework is not overtly apparent, whilst the built fabric remains, sometimes partly de-contextualised, to be interrogated and analysed.

From a tentative start in the early thirties, Cape Town's architecture was to change radically as the concept of modernism was introduced, debated, negotiated, and incorporated into the architecture and planning of the city, so that by 1948 the city appeared to have been changed into a 'modern' city—yet arguably only a simulacrum of the modernist ideal. Modernist architecture was the norm, having entirely supplanted any other architectural form. Yet ironically, the form of the city had merely shifted from a colonial, ethnically divided town to a city whose layout and architecture used modernist principles to reinforce ethnic spatial segregation rather than create a more humane and egalitarian environment—the planning aftermath of which was to endure into the twenty-first century. The political victory of the Nationalist Party in 1948 saw the adoption of modernism, as a mechanism of state; modernist architecture became the visual politics of the state.

Bibliography

Bozdoğan, Sibel. *Modernism and Nation Building. Turkish Architectural Culture in the Early Republic*. Seattle: University of Washington Press, 2001.

Cape Town Housing & Estates Committee minutes. August 30, 1927; May 10, 1928; February 18, 1930.

Cape Town Housing & Slum Clearance Committee minutes. October 11, 1939.

"Cape Town's Coloured Problem." *The Architect, Builder & Engineer*. December, 1935.

Darling, Elizabeth. *Re-forming Britain. Narratives of Modernity before Reconstruction*. London: Routledge, 2007.

Foucault, Michel. *Discipline & Punish: The Birth of the Prison*. Translated by Alan Sheridan. New York: Vintage Books, 1977.

Foucault, Michel. "Of Other Spaces." *Diacritics* 16, no. 1 (1986): 22–27.

Foucault, Michel. *Power*, edited by James Faubion and translated by Robert Hurley and others. London: Penguin Books, 1994.

Foucault, Michel. *The Order of Things. An Archaeology of the Human Sciences*. New York: Vintage Books, 1994.

Foucault, Michel. *The Archaeology of Knowledge*. Translated by Alan Sheridan. London: Routledge, 2004.

Gebhardt, F. "Housing Issues." *The Architect, Builder & Engineer*. January 1936.

Heynen, Hilde. *Architecture and Modernity. A Critique*. Cambridge, Mass: MIT Press, 1999.

Jeppie, Shamil. "Modern Housing in the District: The Canterbury and Bloemhof Flats." In *Recalling Community in Cape Town*, edited by Ciraj Rassool and Sandra Prosalendis, 113–130. Cape Town: District Six Museum, 2001.

Lewis, Desiree. "Writing Hybrid Selves: Richard Rive and Zoë Wicomb." In *Coloured by History, Shaped by Place*, edited by Zimitri Erasmus, 131–159. Cape Town: Kwela Books, 2001.

Mumford, Eric. *The CIAM Discourse on Urbanism 1928*–1960. Cambridge, Mass: MIT Press 2000.

Myers, Garth A. "Intellectual of Empire: Eric Dutton and Hegemony in British Africa." *Annals of the Association of American Geographers* 88, no. 1 (1998): 1–27.

"New Native Township." *The Architect, Builder & Engineer*. October 1923.

Nitzan-Shiftan, Alona. "Contested Zionism-Alternative Modernism: Erich Mendelsohn and the Tel Aviv Chug in Mandate Palestine." *Architectural History* 39 (1996): 147–180.

"Our Slum Problem." *The Architect, Builder & Engineer*. June 1936.

Sangari, Kumkum. "The Politics of the Possible." *Cultural Critique* 7, Fall (1987): 157–186.

"Schotsche Kloof Housing Competition." *The Architect, Builder & Engineer*. April 1939

Scott, James C. *Seeing like a State. How Certain Schemes to Improve the Human Condition Have Failed*. New Haven: Yale University Press, 1998.

Soja, Edward. *Post-Modern Geographies. The Reassertion of Space in Critical Social Theory*. London: Verso, 1989.

Spivak, Gayatri C. 1995. "Can the Subaltern Speak?" In *The Post-Colonial* Studies *Reader*, edited by Bill Ashcroft, Gareth Griffiths and Helen Tiffin, 24–25. London: Routledge, 1995.

"The Urban Problem." *The Architect, Builder & Engineer*. April 1923.

"The White Man's Burden." *The Architect, Builder & Engineer*. June 1927.

Whiteing, Richard. "The Emergence & Development of a Distinct Coloured Identity." MPhil. dissertation, University of Cape Town, South Africa, 2003.

CHAPTER 5

Architectural Agency and 'Place-Making' in a Transformative Post-Apartheid South African Landscape

June Jordaan

Introduction

Around midnight, one month after the opening of the Nyanga Bathhouse in an informal settlement outside Cape Town, vandals started infiltrating the premises. The damage caused was so severe that the building was demolished shortly thereafter. Two years earlier this scheme had been nominated for an Architectural Award of Merit. This example is not idiosyncratic and brings to light that there exist dependencies and presences that lie outside the direct control of the architect. This prompts the question: to what extent do these contingencies impact on the architect's agency of place-making?

For architects, place can be understood as space, created by form, with varying degrees of meaning and significance in time. This chapter suggest that spatial characteristics of place can be grouped into the following categories: tangible versus intangible space; infinite versus finite space; and meaningful and embodied space.

The ambiguity of architectural agency is a concern that exists in both the practice and training of architects. This chapter aims to provide insight into the agency of the architect in 'place-making' in post-apartheid South Africa. This will be done by explicating various post-apartheid places through a set of spatial theories. By making reference to these existing spatial theories, this chapter will develop a spatial triad and apply it to a unique South African place, the Cape Town Railway Station. This building has been subject to two major spatial interventions over time, or, it could be argued, subject to two attempts at architectural place-making. The earlier occurred in a late colonial and apartheid context, in which South African modernist architecture was appropriated as 'a mechanism of power and control'[1] and the latter, in a democratic and

1 André van Graan, "Modernism as a Mechanism of Power and Control in Colonial Contexts: The Project of Modernity in Cape Town South Africa," in this volume.

 | DOI 10.1163/9789004376793_006

transformative post-apartheid context. Moreover, this study hopes to illustrate how a multidimensional approach to place, that incorporates time, may explicate dependencies and presences that lie outside the realm of the architect.

Autonomous Agents in Transformative Contexts

It suffices to say, that as architects we should be trained to be free thinkers. This training, however, often happens in a self-referential world, detached from reality and time. In this 'deluded detachment', we are often made to think that we have complete autonomy in our practice of designing meaningful and functional environments.[2] Our studio spaces, physically and metaphorically removed from the outside world, and the rituals and practices within these spaces, reinforce this delusion. Banham argues that these strange characteristics and rituals may lead to the marginalization and potential distinction of architecture as a discipline.[3] What is essentially wrong with this scenario for architects is that it isolates the idea of space with that of time. At the beginning of the 20th century, the theory of relativity, or the space-time continuum, was embraced by various artists and architects.[4] The theory of relativity discards the premise of absolute time and space and that moving systems are dependent on these.[5] Instead, it sets forth that time and space are not absolute and that they are relative to the moving systems. Time and space are therefore not vessels for the moving bodies of the universe, but can in fact not exist without them.[6] Similarly, Lefebvre stresses the importance of considering time

2 Jeremy Till, *Architecture Depends* (Massachusetts: MIT Press, 2009), 7. Till refers to a literal and symbolic detachment that architecture schools have from reality or 'place'. The detachment relates to the fact that, the world is not seen as a dynamic social system to be engaged with and which is open to transformation, but instead as a static abstraction, existing only to receive mute form.

3 Reyner Banham, "A Black Box: The Secret Profession of Architecture", in *A Critic Writes: Selected Essays by Reyner Banham*, eds. Mary Banham, Sutherland Lyall, Cedric Price, and Paul Barker (Berkeley: University of California Press, 1996), 295.

4 Cornelis van de Ven, *Space in Architecture: The Evolution of a New Idea in the Theory and History of the Modern Movements* (Michigan: Van Gorcum, 1978). Here, the author discusses how architects such as Theo van Doesberg, Walter Gropius, and Sigfried Giedion conceived of time and space as part and parcel.

5 Ibid.

6 Ibid.

with space. He argues that (social) space is a (social) product.[7] Here, space is imagined not as an independent material reality existing 'in itself'. Instead it posits space as fundamentally bound up with reality.[8]

The concern that architects tend to design places in denial of reality, conforming purely to their elitist, self-referential discipline, has been raised by many. Till for example highlights how people refer to RIBA, the Royal Institute of British Architects, as '*Remember, I'm the bloody Architect!*'[9] This concern is as relevant in South Africa as anywhere else. The Nyanga Bathouse, for example, located in an informal settlement outside Cape Town, was based on the so-called stock tools of 'place-making', and was nominated for an architectural award of merit.[10] Two months after opening, however, it had been turned into a slaughter house for cows and was shortly thereafter vandalized severely and levelled with the ground. Other places have been re-appropriated in more positive ways. The Voortrekker Monument, an icon of the 'Afrikaner' culture and arguably with apartheid, has recently been used as a rock music venue. Drawing on the 'zef' counter-culture movement, the South African rap-rave music group, Die Antwoord, also shoot music videos in places that were previously regarded as slums.

The relevance of place-making in South Africa has been highlighted by many. In South Africa, 'place' is an ideologically-loaded word that has often been referred to as a 'healing instrument' through which the excesses and abstraction of apartheid's modernist space might be redressed.[11] The universal and universalizing 'space' of modernism was regarded by some as a culprit, which generates disaffection in society.[12] Built environments from the apartheid years to a large extent still contribute to these geographies. Current

---

7 Henry Lefebvre, *The Production of Space*, trans. Donald Nicholson-Smith (Oxford: Basil Blackwell, 1991).

8 Christian Schmid, "Henri Lefebvre's Theory of the Production of Space: Towards a Three-Dimensional Dialectic," in *Space, Difference and Everyday Life*, eds. Kanishka Goonewardena, Stefan Kipfer, Richard Milgrom, and Christian Schmid (New York: Routledge, 2008), 28.

9 Till, *Architecture Depends*, 153.

10 Nic Coetzer, "Exploring 'Place-making', City Squares and Other Places: Cape Town's Pre-apartheid Spatial Politics." *South African Journal for Art History* 23, no. 1. (2008): 139. Here, Coetzer refers to courtyard buildings and city squares as stock tools of place-making proponents.

11 Ibid., 140.

12 Ibid.

'place-making' strategies in South Africa are often concerned with the question of a post-apartheid South African identity, and have been employed in the architectural design of public spaces since the early 1990s.[13] The Constitutional Court Precinct, Freedom Park, Walter Sisulu Square, Nobel Square and various other judicial precincts exemplify post-apartheid place-making in South Africa, where architects sought to interpret the meaning and significance of particular locales (figure 5.1). Different design approaches exist in this regard: Some could be seen to refer to the past, others to rupture from the past, and others as hybrids. Nearly thirty years after apartheid, South Africa is still in a process of transition, and place-making—or in cultural geographers' terms, the creation of meaningful and significant spaces—forms an integral part of it.

FIGURE 5.1 *Post-apartheid place-making in South Africa has proliferated in the design of new public spaces, such as Nobel Square in Cape Town*
PHOTO: JUNE JORDAAN.

13 Jonathan Noble, "White Skin, Black Masks: On Questions of African Identity in Post-Apartheid Public Architectural Design 1994–2006" (PhD diss., University College of London, UK, 2007). Also see Lindsay J. Bremner, "Memory, Nation Building and the Post-Apartheid City," in *Desire Lines: Space, Memory and Identity in the Post-apartheid City,* eds. Noëleen Murray, Nick Shepherd, and Martin Hall (New York: Routledge, 2007), 85–103.

Finite, Tangible Space

The opening chapters of *Tao Te Ching*, a fundamental text for Taoism, reveal the superiority of the contained, the space within.[14] Hereby, the non-existent, made tangible in material form, becomes essential.[15] This conceptualization of space alludes to the ancient oriental saying: 'mass is the servant of the void'. Similarly, Aristotle proposed a conceptualization of space, called place or *topos*. Place in the Aristotelian sense implies a fixed boundary. Despite the discovery of the ever-extending infinite cosmos, this conceptualization of space has been embraced by many. This 'finite space' was underpinned by phenomenological contributions that question the relevance of the scientific knowledge of the universe, since the concretely experienced space to be lived in, has a character of a finite, enclosed interior.[16]

Similarly, Norberg-Schulz argues that a boundary is not that point where something stops but, as the Greeks recognized, the boundary is that point where something begins its 'presencing'.[17] Norberg-Schulz categorizes inside and outside by their natural and physical barriers. For him *genius loci* can be achieved through materials and construction. Similarly, Lynch analyses elements that contribute to legibility of a place, which indirectly, at least defines a sense of place to be 'things' with 'character' and 'meaning'.[18] Our environments therefore should be carved up into clearly defined, distinctive nodes that are easily differentiated from their background.[19] From these authors it is easy to surmise that the role of the architect is to provide good 'form' wherein communities can be stabilized, dwelling can take place, and thereby produce happy, non-alienated, non-conflicted human beings.[20] South African places that are physically bounded or at least clearly defined with strong identifiable forms are manifold, but often controlled, manicured, and 'disneyfied' environments. Nelson Mandela Square, for example, is a nostalgic and superficial reproduction of the medieval European city square.

In Cape Town, numerous places that have been designed to memorialize the ills of apartheid are controlled environments. In Freedom Park, for example,

14 Lao Tzu, *Tao Te Ching*, trans. Charles Muller (New York: Barnes & Noble, 2005).

15 Van de Ven, *Space in Architecture*, 3.

16 Ibid.

17 Christian Norberg-Schulz, *Genius Loci: Towards a Phenomenology of Architecture* (London: Academy Editions, 1980), 13.

18 Kevin Lynch, *The Image of the City* (Cambridge MA: MIT Press, 1960).

19 Coetzer, *Exploring 'Place-making'*, 141.

20 Ibid.

one must be escorted throughout, and is informed on what design elements signify. The Apartheid Museum and the Hector Peterson Museum are also environments that are dependent on systems that keep the 'unwanted' out, or environments where one might be, as Cresswell asserts, 'in place' but feel 'out of place'.[21] Local architectural interventions in environments that are bounded or clearly distinguishable often 'freeze time'.[22] The Bo-Kaap housing area for example, with its Cape-Malay origins, is distinct with its brightly colored facades. Architectural interventions here cannot be recognized as new buildings are built as exact replicas of the older buildings. Similarly, the redevelopment plans of District Six, an area where mainly colored people lived and which was demolished and declared a 'white area' under the apartheid government, suggest an environment similar to that which was fine grain urbanism with small street blocks and narrow roads.

Environments resulting from this kind of architectural approach is visually seductive but also nostalgic and often kitsch. Here, architects have agency, but only if these environments are controlled. This approach is to a large extent in denial of time: time is in actual fact 'frozen' in order to capture a state of perfection.[23] Identifiable environments are created but they are often exclusive and may also create alienation. A concern in the South African context that cannot be over-estimated is the necessity for the exclusion of crime and the thus the 'unwanted'.

An Open, Progressive Sense of Place

Spaces that are not bounded, clearly identifiable, or distinguishable from their environments have often been referred to as 'non-places' or 'thinned out' places. With respect to 'non-places', literary and architectural critics have argued that buildings such as airports can indeed have a sense of place, not the nostalgic kind but related to the camaraderie.[24] The film *The Terminal* by Steven Spielberg (2004) adequately illustrates this point. In this instance the airport, a quintessential 'non-place' becomes the alienated person's home, a

21 Tim Cresswell, *In Place/Out of Place: Geography, Ideology, and Transgression* (Minneapolis: University of Minnesota Press, 1996).

22 Till, *Architecture Depends*, 181. Here Till argues that architects often banish time from their work, sometimes freezing it and other times sidestepping it altogether.

23 Ibid.

24 Pattabi G. Raman, *Between Typological Rigidity and Flight of Fancy: Cape Town Airport* (Cape Town: Picasso, 2012), 2.

quintessential 'place'.[25] If we follow these lines of enquiry perhaps we can see in modernist public space, despite its bleakness, a certain desire to move away from the totally enclosed and sometimes claustrophobic pre-industrial space to expansive 21st century spaces with plenty of light and air and more importantly, spaces that are subtly and possibly invisibly connected to the rest of the city. For example, the Walter Sisulu Square in Soweto may be interpreted as being open and connected. Perhaps this is what Mondrian implied in his paintings of *Place de la Concord* (1943) and *New York* (1944). Does the idea of modernism as an 'incomplete project' maintain certain potential for enlightenment?[26] Casey argues that chaos may very well be a scene of emerging order.[27] Could it be that, in the present world, what we see as chaos is in fact superior order? Similarly, Massey suggests a progressive sense of place that is open, hybrid, multiple, contested, fluid, and uncertain, rather than fixed territorial units.[28]

In South Africa, these arguments have tremendous merit. Places in developing parts of the country exemplify an incredible sense of human ingenuity, an ability to change and adapt to individual needs. These processes are often organic and exhibit a true spirit of invention (figure 5.2). Here Thrift's 'non-representational' nature of place is of relevance: The focus is on 'events and practices' rather than 'interpretation and representation'.[29] Arguing that encounters with place cannot be adequately registered through language or discourse, he stresses the importance of how we intuitively inhabit spaces. Thereby place is created through our 'rhythm of being'. The way we intuitively inhabit places may also be seen as the action of imbuing place with meaning. This is in line with Seamon's 'place-ballet'.[30] Hereby meaning could be regarded as fragmented and ambivalent and architecture may be seen as 'making place' for an event.[31]

What becomes evident in the distinction between the two architectural approaches discussed above is that the former is subject to controlled

25 Bruce Janz, "Coming to Place," *Environmental and Architectural Phenomenology Newsletter* Fall 15, no. 3 (2004).

26 Jürgen Habermas and Seyla Ben-Habib, "Modernity vs Post Modernity," in *New German Critique* 22 (1981): 3–14.

27 Edward Casey, *The Fate of Place: A Philosophical History* (Berkeley: University of California Press, 1998).

28 Doreen Massey, "A Global Sense of Place," in *Reading Human Geography*, eds. Trevor Barnes and Derek Gregory (London: Arnold, 1997), 315–323.

29 Nigel Thrift, *Non-Representational Theory: Space, Politic, Affects* (London: Sage, 2007).

30 David Seamon, *Geography of the Lifeworld: Movement, Rest and Encounter* (New York: St Martin's Press, 1979).

31 Jacques Derrida, "Point de Folie-Maintenant L'Architecture," trans. Kate Linker, *AA Files* no. 12 (1986): 65–75.

FIGURE 5.2 *The non-representational nature of space exemplified through spontaneous place-making. Here, graffiti artist Faith47 depicts the harsh realities of post-apartheid urban living, by contrasting emancipatory phrases from the South African Freedom Charter with urban realities of ruination and decay*
PHOTO: JUNE JORDAAN.

environments and to a large extent in denial of the contingencies that are brought about by time. The meaning of these spaces is also understood as being fixed. The latter inclines to environments that facilitate time and the contingencies it implies. These environments are non-representational, and therefore facilitate change but run the risk of being merely functional and imaginatively and emotively deprived (table 5.1).

In South Africa, there exist architectural projects that exhibit qualities of both approaches. These projects, identified as 'post-apartheid hybrids', are more or less generic spatially, but have additional layering of local motifs. Projects such as the Baragwanath Transport Interchange and the Constitutional Court have the inclusion of craft, tectonic, and sculptural elements as tropes of a formerly denied history.[32] This layering is most commonly material and not spatial. Facades of these buildings are thickened with decorative motifs that are intended to open dialogues with different narratives.[33] These projects have

32 Jonathan Noble, *Architecture, Hybridity, and Post-Apartheid Design* (Johannesburg: Art Historical Work Group, 2008), 71.

33 Ibid.

TABLE 5.1 *Opposing theories in 'place-making'.*

Finite, bounded, differentiated space	Infinite, open, fluid space
Manicured	Allows for spontaneity, dirt, chaos
Pastiche + disneyfied	Allows for human ingenuity + change
Exclusive + controlled	Inclusive + subject to crime
Big narratives	Small narratives
Identifiable character	Non-representational character
Meaning fixed	Meaning ambivalent
Hybrids	Incomplete modernist project

had a mixed reception. Raman for example argues that: 'when contemporary architecture directly uses indigenous motives, it verges on the kitsch; when it aims to abstract ideas form the African past, it ends up mocking it.'[34] Some other projects build on the past instead of superficially adding it. This approach resonates with Habermas's notion of the unfinished modernist project. It argues that tradition can be blind and implies that the principles of modernism are universal and not a style.[35]

The theories discussed above have merits and shortcomings and their limitations and contradictions in the current South African context have been highlighted. Hybrids are developing, but are still to a large extent material and not spatial. In South African context, architectural projects that allow for participation and the evolution of meaning are occurring and have a firmer reciprocity with reality. A valuable way to discern between the approaches discussed is to take cognisance of the fact that they are all, to a greater or lesser degree, concerned with the way that people experience and embody spaces. Indeed, one thing that does seem to be widely agreed upon is that place has to do with embodiment, that place is something irreducible and essential to being human.[36]

---

34 Pattabi G. Raman, "Peter Rich: Incomplete Modernist Project Versus Robust Hybridity in South," *Architecture+* 3 (2003): 108.

35 Ibid. Peter Rich, for example feels that Ndebele settlements are exemplary in embracing much of modernism's principles of asymmetry, spatial flow, functional efficacy, and minimalist expression, but is as colorful as the latter day rainbow modernism of Also van Eyck.

36 Martin Heidegger, *Being and Time* (New York: Harper Perennial Modern Classics, 2008); Robert D. Sack, *Place, Consumption and Modernity* (Baltimore: Johns Hopkins University Press, 1992); Robert D. Sack, *Homo Geographicus* (Baltimore: Johns Hopkins University Press, 1997); Jeff Malpas, *Place and Experience: A Philosophical Topography* (Cambridge:

The Material, Lived, and Mental Dimensions of the Lifeworld

Numerous scholars, such as Lefebvre and Soja, have developed valuable conceptual triads that incorporate the body with space and time.[37] It has been noted that the contemporary relevance of these conceptualizations is strengthened by globalization and urbanization.[38] The current reception of these triads, however, has not made full use of their possibilities, particularly not by the discipline of architecture. By drawing on these theories, this study suggests a simplified version of a conceptual triad in the form of three spatial dimensions of the lifeworld: the material, lived, and mental. These spatial dimensions, similar to the triads above, are conceptualized to exist in an ongoing state of reciprocity. This is in line with Thijs-Evenson's argument that the physical, material world, and human experience are part and parcel: through our sensory, moving bodies, we experience physical forms in varying degrees.[39] By bracketing private, social, and universal experiences, these forms can be identified and designed to have certain archetypical expressions.

Firstly, *the material* designates the physical and tangible elements that constitute places. This visual, but also tactile dimension of place has the ability to seduce movement and has expressive qualities. As a result of its tangible nature, this spatial dimension receives great emphasis in architectural teaching. It has also been criticized as it permits ocularcentric modes of design and abstractions of reality, often generated intellectually instead of emotively.[40] This dimension can be probed hermeneutically.

Secondly, *the lived* dimension is concerned with the way in which place as a phenomenon is not static, but instead something that's meaning and significance is in a constant state of flux. Places therefore never 'are' but are always 'becoming'. Furthermore, it is concerned with how places are subject to the way people interact with them, how our rhythms, rituals, and events in places influence the ways in which they are experienced. The disciplines of environmental behavior and cultural geography provide valuable insights here, such as Seamon's place-ballet, Ingold's work on path making, and Lynch's study on

---

Cambridge University Press, 1999); Casey, *The Fate of Place*; Maurice Merleau-Ponty, *Phenomenology of Perception* (London: Routledge, 1962).

37 Lefebvre, *The Production of Space*. Also see Edward Soja, *Thirdspace: Journeys to Los Angeles and Other Real-and-Imagined Places* (Oxford: Blackwell Publishers, 2000).

38 Schmid, *Henri Levebvre's Theory*, 27.

39 Thomas Thijs-Evenson, *Archetypes in Architecture* (New York: Oxford University Press, 1989).

40 Joseph Dean, "The Great Divide," *The Architectural Review* 233, no. 1392 (2013): 18–19.

the legibility of the city.[41] By referring to bodily actions, this dimension sheds light on supposed normative behavior and transgression in particular places.

Thirdly, *feeling* denotes the way in which our experiences of places are produced through memory, imagination, and dreams. This dimension of place resides in our mind. It is less tangible than the other dimensions, as it cannot be explicated through interpretation of physical characteristics or patterns of use, but could arguably be seen to have the strongest force in our experience of place. By referring to examples in the arts, i.e. pictorial, fictional, and cinematic place, this dimension can show us how places of memories, imagination, and also dreams are liberated from the utilitarian constraints and can therefore provide valuable insights into subtle and responsive place-making.[42] The Cape Town railway station is a significant place in South Africa's current post-apartheid context that has undergone various changes though time, and that can be interpreted through these dimensions.

The Evolution of a Place

The Cape Town railway station has a long and rich history. Throughout time its material qualities, utilitarian function, and meaning has undergone significant changes. Two major interventions that affected the evolution of this place can be seen in its transition from a colonial to a modernist building during apartheid, and thereafter in its transition to a post-apartheid, transformative, and inclusive place.

The first formal station structure on this site was a grandiose Victorian building, a product of colonialism. During the height of apartheid, this 'old lady' was demolished and a high modernist building put in its place. This intervention did not only facilitate segregated commuting, but was also a grand narrative,[43] employed by the government that signified apartheid's domination and oppression. This building can be referred to as 'an oppressor', as it occurred under the rule of apartheid and concretized its political ideologies. The political climate of this time, and its effect on the material and psychological

41 Seamon, *Geography of the Lifeworld*; Lynch, *The Image of the City*; Tim Ingold, *The Perception of the Environment: Essays on Livelihood, Dwelling and Skill* (London: Routledge, 2006); Lynch, *The Image of the City*.

42 Juhani Pallasmaa, *Architecture of Image: Existential Space in Cinema* (Helsinki: Rakennustieto, 2007).

43 Jean-François Lyotard, *The Postmodern Condition: A Report on Knowledge* (Minneapolis: University of Minnesota Press, 1984).

geographies—its effects on its places—has been elaborated on by van Graan.[44] In his chapter in this volume, he stresses that in this context, modernism was at odds with the utopian socialist concern for improving the well-being of its entire people, and instead became a mechanism for 'improving production and restructuring urban areas into racially and physically segregated areas'.[45] He notes that civic authorities experimented with modernism in architecture for the indigenous population, the 'Colonial Other' that was completely disempowered and had no say in the design process.[46] These arguments can certainly be seen in the programmatic and utilitarian design of this station. Here, the access and movement of the oppressed members of society was strictly controlled in and around the building, and regulated through designated 'non-white' areas (figure 5.3). As a result of the global 'project of modernism' we can safely assume that this building's forms have not been conceptualized as vernacular responses or entities with existential expressions, but instead as local adaptations of global architectural movements. Furthermore, it can be assumed that this building's exclusive and controlling nature left little room for any human ingenuity, spontaneity, or change.

In recent years this station was subject to a further architectural intervention, to another attempt in place-making. This latest design can be regarded as 'the equalizer' as it occurred after the demise of apartheid and sought to undermine the spatial ideologies of the previous regime (figure 5.4). This

FIGURE 5.3 *The exterior of the current Cape Town Railway Station still exhibits the high modernist architecture of the apartheid station*
PHOTO: JUNE JORDAAN.

44 van Graan, "Modernism as a Mechanism of Power and Control in Colonial Contexts."

45 Ibid.

46 Ibid.

intervention can also be seen as a set of smaller narratives as it consists of a series of small interior interventions that undo the forced segregated commuting and that contribute to fresh, open, and connected interior environments, even possibly to a progressive sense of place. The new complex was designed and timed to be open and be fully functional for South Africa's hosting of the 2010 World Cup Soccer.

The new areas within and around the station, despite its perceived rupture from its past, are to large degree still controlled and do not allow for spontaneous urbanism.[47] Spaces around the station have no doubt been gentrified. The presence of crime and the 'unwanted' are realities that persist. In this latest intervention, a light box was added to the roof of the station and a new skin constructed to the Strand Street facade, which contributes, to a small extent, to the legibility and identification of the new station. Although newness, freshness, and openness are experienced inside the building, the exterior still overtly expresses the stale mood of apartheid's 'oppressor'. Some physical remnants of the preceding interior were kept intact, such as murals indicating the Afrikaner's 'Groot Trek', an event in South Africa's history that relates to the liberation of the Afrikaner from the British Colony in the Cape. This evidence of 'times gone by' that draws on collective memory and association, creates a palimpsest and contributes to the temporal nature of the new interior. Heritage regulations, of course, played a major role in determining to what extent such historical traces were preserved and discarded.

Informal trading that used to occur in the forecourt was formalized and moved to the roof of the station, an area still mostly used by the previously disadvantaged. This move has been to the detriment of small businesses. Financial forces allowed major international franchises such as McDonalds and KFC to be located in high pedestrian traffic areas, undoubtedly compromising the identity and spirit of station as a node in the rich colonial Cape Town context.

Positive platial repercussions of this intervention can certainly be noted, particularly on a broader scale. The new station was done within a comprehensive and extremely restrictive urban design framework. The urban framework in which the station sits connects it to the new Cape Town soccer stadium through the city by means of the 'freedom fan-mile'. Along this walk there are spaces that reveal inclusively and exclusivity. It is filled with heterotopias that are in flux. How people intuitively inhabit the fan-mile are on-going factors that impact on the meaning and significance of its spaces. The stadium attracts people from previous non-white 'locations'. The fan-mile facilitates events

47 Raman, email to author 2012. Referred to this occurrence as people spontaneously and intuitively inhabiting space, such as skateboarders, graffiti artists, and protestors.

FIGURE 5.4 *The segregation of commuters has been undone within the interior of the Cape Town Railway Station*
PHOTO: JUNE JORDAAN.

that during apartheid undoubtedly constituted spatial transgression, such as gay-pride parades, naked bike rides, and the all-inclusive Cape Town carnival. These naked bike rides, that occur once a year on the day before the major Argus Cycling challenge, commences and ends in Prestwich Place (figure 5.5). This locale has for a long time been subject to class and racial conflict. It was the site of mass graves of the Dutch Reformed church and also of colonial slaves. Just over a decade ago, construction on this site led to the uncovering of a significant amount of human bones due to which development of the area was halted and the site turned into a memorial.

By looking at these examples various contingencies can be highlighted that impact on architectural agency in place-making. In the case of the first transition two major factors can be noted. On the one hand, political ideology and the power of civic authorities undoubtedly played a large role. On the other hand, the architectural meta-narrative of the 'modernist project' was a major drive. These two forces, the one ideological and thus intangible, and the other architectural and thus tangible, strengthened one another and affected a framework in which the architects undoubtedly had limited agency.

In the case of the second transition the new political ideology, i.e. democracy, can certainly be identified as a major contributing factor that influenced

FIGURE 5.5 *The annual naked bike commencing at the Prestwich Memorial, a locale in which mass graves of the Dutch Reformed Church and slaves from colonial times were recently uncovered*
PHOTO: JUNE JORDAAN.

place-making at the station. Financial restrictions and control, the fear of crime and the 'unwanted', heritage regulations, and the restrictive nature of a larger planning scheme, such as the urban design framework, can also be noted to have stifled architectural agency.

Probing the contingencies and presences that lie outside our scope of work allows us to look at architectural places beyond face value. It enables us to identify the diverse forces that impact on our agency, but it also allows us to look at the repercussions of places, repercussions over which we have no control. In this regard, the first transition discussed above, from the colonial to the modernist apartheid building, exemplifies what Foucault designates as being central to power relations: to render visible both the mechanism for the wielding of power and to create mechanisms of surveillance through architecture and planning as a means of control.[48] The architecture of this modernist station could therefore be seen as 'a mirage that obscured its real motives'.

In the case of the second transition, these themes certainly also become relevant. Positive repercussions of this place can be seen in both the progressive places that have manifested, and events that occur in and around the station,

48 Michel Foucault, "Of Other Spaces: Utopias and Heterotopias," *Architecture /Mouvement/ Continuité*, translated by Jay Miskowiec, 5 (1984): 46–49.

particularly along the freedom fan-mile. There are however, also sinister platial consequences to be noted. With the launch of the scheme, during the time of the Soccer World Cup, the unwanted vagrants living around the station, were picked up and moved to an artificial settlement named Blikkiesdorp. In Afrikaans Blikkiesdorp means 'Tin Town'. It is a planned tin shack town that is located far outside the city and referred to by its residents as an 'apartheid dumping ground' or a 'concentration camp'. This place is inextricably, albeit not visibly, linked to this station, through time, events, and memories. The material, lived, and mental aspects of this place is in stark contrast to other open, free, connected, and heterogeneous spaces associated with the new station. The physical forms that constitute this place and their expressive qualities can certainly also be condemned. It disturbingly resembles the memorial to the murdered Jews in Berlin, as a repetitive, monotonous and homogenous landscape (figure 5.6). It can safely be assumed that the types of experiences that are fundamentally at odds with 'places', such as disorientation, alienation, and existential angst, will be felt here.

The undoing of segregation, exclusivity, and oppression in this last transition of the station can undoubtedly be challenged when taking these broader

FIGURE 5.6 *A geography of nowhere, the memorial to the murdered Jews in Berlin unnervingly resembles the residential landscape of Blikkiesdorp*
PHOTO: JUNE JORDAAN.

spatial consequences into consideration. The architect, however, cannot possibly be held responsible, as the circumstances that framed the architectural production, or in the case of this study, the 'platial production', were forces of undeniable magnitudes. Indeed, the architects in both instances cannot be held fully accountable as they never had full autonomy in effecting these places. This point is echoed by Till who contests the idea that architectural progress is affected by previous architectural movements.[49] In this regard, architects are not viewed as the inventors of movements, but instead as inevitable consequences thereof,[50] a point that is in line with Marx's view that 'men make history but not in circumstances of their own choosing'.

The dark shadow of Blikkiesdorp is still around. Many of the vagrants who have not made their way back to the city still calls this place home. This shadow is not idiosyncratic to the Cape Town station and should be a reminder that the creation of meaningful and embodied environments is complex and influenced from many diverse and often unforeseen angles. Furthermore, the repercussions of places are often unpredictable and even invisible. It proves important to note however, that the effects of such places on the everyday experienced lifeworld cannot be underestimated. The space in which architects are able to manoeuvre to control these effects, remains limited.

Gymnastics in the Prison Yard

> How ineffectual are the brilliant gymnastics carried out in the yard of the model prison, in which architects are free to move about in temporary reprieve?[51]

Tafuri expresses the concern that architecture, caught between the structures of capitalism, has lost any means of resistance. He also states that architecture has deluded itself into believing that the production of form alone can intervene productively in the social world.[52] This chapter has indicated by means of various architectural examples that these concerns have merit in the context of post-apartheid architectural agency. Architects have referred to various spatial theories to generate forms in their endeavor of 'place-making'. Such theories are often cerebral and self-referential to our discipline and disregard the circumstances that frame 'platial production'.

49 Till, *Architecture Depends*, 33.

50 Ibid.

51 Manfredo Tafuri, *Theories and History of Architecture* (London: Granada, 1980), xxii.

52 Ibid.

In this chapter, a number of these theories have been discussed in relation to current post-apartheid South African cases. Through this it has been shown that public places, unless rigidly controlled, inevitably result in unpredictable environments. It was also shown that there exists, amongst others, static, fluctuating, and hybrid approaches to place-making. Furthermore, this study indicates that there exists external informants and unpredictable repercussions of place-making that are often not overtly visible to our profession. A valuable way to discern between the merits of the approaches discussed above is to take cognisance of the fact that they are all, to a greater or lesser degree, concerned with the way that people experience and embody spaces.

To better understand these concerns priority should therefore be given to the experiential and embodied dimensions of place. These denote the tangible, but also the intangible aspects of places, the visible and the invisible. Places should therefore not merely be considered at face value but instead as phenomena that affect our lifeworlds in various ways and on various scales. Places are thus ultimately tied up with the way people use them and the meaning they attach to them. Innovative examples that take cognisance of this are schemes that allow for flux, user participation, and incremental change.

As architectural training is where most of the values of the profession are established, it becomes important to consider how we can help students understand these points, particularly in terms of spatial agency. To incorporate the phenomenological body into an understanding of place, conceptual devices should be explored—devices that are simplified and can be explained to students, devices that can be applied to cases to explicate architectural agency and provide insight into contingencies and presences that lie outside the scope of the architect. Three dimensions of the lifeworld has been suggested in this regard, to work as a conceptual device in which to interpret and postulate place. These consist of the material, the lived, and the mental dimensions of place. Making reference to these dimensions might help us recognize that we do not live separately in material worlds and mental worlds, but that these experiential dimensions are fully intertwined. Furthermore, it might help us understand how architects are able to contribute positively to these experiential dimensions of place, and to what extent.

Bibliography

Banham, Reyner. "A Black Box: The Secret Profession of Architecture." In *A Critic Writes: Selected Essays by Reyner Banham*, edited by Mary Banham, Sutherland Lyall, Cedric Price, and Paul Barker, 292–299. Berkeley: University of California Press, 1996.

Bremner, Lindsay J. "Memory, Nation Building and the Post-Apartheid City." In *Desire Lines: Space, Memory and Identity in the Post-apartheid City*, edited by Noëleen Murray, Nick Shepherd, and Martin Hall, 85–103. New York: Routeledge. 2007.

Casey, Edward. *The Fate of Place: A Philosophical History.* Berkeley: University of California Press, 1998.

Coetzer, Nic. "Exploring 'Place-making', City Squares and Other Places: Cape Town's Pre-apartheid Spatial Politics." *South African Journal for Art History* 23, no. 1. (2008): 139–157.

Cresswell, Tim. *In Place/Out of Place: Geography, Ideology, and Transgression.* Minneapolis: University of Minnesota Press, 1996.

Cresswell, Tim. *Place: A Short Introduction*. Oxford: Blackwell, 2005.

Dean, Joseph. "The Great Divide." *Architectural Review* 233, no. 1392 (2013): 18–19.

Derrida, Jacques. "Point de Folie-Maintenant L'Architecture." Translated by Kate Linker. *AA Files* no. 12 (1986): 65–75.

Foucault, Michel. "Of Other Spaces: Utopias and Heterotopias." Translated by Jay Miskowiec. *Architecture /Mouvement/ Continuité* 5 (1984): 46–49.

Habermas, Jürgen and Seyla Ben-Habib. "Modernity vs Post Modernity." *New German Critique* 22 (1981): 3–14.

Habraken, John. *Palladio's Children*. London: Taylor and Francis, 2005.

Heidegger, Martin. *Being and Time*. London: Wiley-Blackwell, 1967.

Ingold, Tim. *The Perception of the Environment: Essays on Livelihood, Dwelling and Skill.* London: Routledge, 2006.

Janz, Bruno. "Coming to Place." *Environmental and Architectural Phenomenology Newsletter* Fall 15, no. 3 (2004). Accessed May 18, 2013. http://www.arch.ksu.edu/seamon/janz.html.

Lefebvre, Henri. *The Production of space.* Translated by Donald Nicholson-Smith. Oxford: Basil Blackwell, 1991.

Lynch, Kevin. *The Image of the City*. Cambridge MA: MIT Press, 1960.

Lyotard, Jean-François. *The Postmodern Condition: A Report on Knowledge*. Minneapolis: University of Minnesota Press, 1984.

Malpas, Jeff. *Place and Experience: A Philosophical Topography*. Cambridge: Cambridge University Press, 1999.

Massey, Doreen. "A Global Sense of Place." In *Reading Human Geography*, edited by Trevor Barnes and Derek Gregory, 315–323. London: Arnold, 1997.

Massey, Doreen. *For Space*. London: Sage, 2005.

Merleau-Ponty, Maurice. *Phenomenology of Perception*. London: Routledge, 1962.

Noble, Jonathan. "White Skin, Black Masks: On Questions of African Identity in Post-Apartheid Public Architectural Design, 1994–2006." PhD Dissertation, University College of London, UK, 2007.

Noble, Jonathan. "Architecture, Hybridity, and Post-Apartheid Design." *South African Journal for Art History* 23, no. 2 (2008): 71–88.

Norberg-Schulz, Christian. *Genius Loci: Towards a Phenomenology of Architecture*. London: Academy Editions, 1980.

Pallasmaa, Juhani. *The Eyes of the Skin: Architecture of the Senses.* London: Academy Editions, 1996.

Pallasmaa, Juhani. *Architecture of Image: Existential Space in Cinema*. Helsinki: Rakennustieto, 2007.

Raman, Pattabi G. "Peter Rich: Incomplete Modernist Project versus Robust Hybridity in South Africa." *Architecture+* 3 (2003): 108.

Raman, Pattabi G. *Between Typological Rigidity and Flight of Fancy: Cape Town Airport.* Cape Town: Picasso, 2012.

Raman, Pattabi G. Email to author, 21 June 2012.

Rodaway, Paul. *Sensuous Geographies: Body, Sense and Place*. London: Routledge, 1994.

Sack, Robert David. *Place, Consumption and Modernity*. Baltimore: Johns Hopkins University Press, 1992.

Sack, Robert David. *Homo Geographicus*. Baltimore: Johns Hopkins University Press, 1997.

Schmid, Christian. "Henri Lefebvre's Theory of the Production of Space: Towards a Three-Dimensional Dialectic." In *Space, Difference and Everyday Life*, edited by Kanishka Goonewardena, Stefan Kipfer, Richard Milgrom, and Christian Schmid, 27–46. New York: Routledge, 2008.

Seamon, David. *Geography of the Lifeworld: Movement, Rest, and Encounter*. New York: St Martin's Press, 1979.

Soja, Edward. *Thirdspace: Journeys to Los Angeles and Other Real-and-Imagined Places.* Oxford: Blackwell Publishers, 2000.

Tafuri, Manfedo. *Theories and History of Architecture*. London: Granada, 1980.

Thijs-Evenson, Thomas. *Archetypes in Architecture*. New York: Oxford University Press, 1989.

Thrift, Nigel. *Non-Representational Theory: Space, Politics, Affect*. London: Sage, 2007.

Till, Jeremy. *Architecture Depends*. Massachusetts: MIT Press, 2009.

Tzu, Lao. *Tao Te Ching*. Translated by Charles Muller. New York: Barnes & Noble, 2005.

Van de Ven, Cornelis. *Space in Architecture: The Evolution of a New Idea in the Theory and History of the Modern Movements*. Michigan: Van Gorcum, 1978.

CHAPTER 6

Women as Retrofits in Modernist Low-Income Housing

Assumpta Nnaggenda-Musana and Eiman Ahmed Elwidaa

Charting Colonial Planning in Uganda

Planning in Uganda's urban areas developed on the basis of British colonial urban planning influences and ideologies.[1] Mukwaya notes that 'the Colonial period defined Uganda's spatial and urban development pattern.'[2] Under colonial planning Uganda was broken down into provinces, which were sub-divided further into districts. The headquarters of these provinces and districts were later to become urban centers and the first townships in Uganda.[3] Some of Uganda's large urban areas are Kampala and Jinja, which are the focus of this chapter.

African housing during the colonial times was not designed for female habitation. However, architects should recognize that the creation and utilization of space both shapes and is shaped by social relations. Houses should be recognized as spheres of feminine activity since women stay longer at home. Have women been considered in post-colonial housing? Integrating women's needs in housing can make their lives easier for collaborative and productive income-generating activity while permitting income accumulation and time for other activities. The empirical part of this chapter explores the adequacy of the government provided houses for women in the MWHS: Masese Women's Housing Scheme in Jinja. Findings were generated from field research, archival data, and oral histories to reveal the women's ways of living. Employing a qualitative approach enabled the researchers to ground their analysis of the women's interpretation of real life. The study attempts to understand gendered social relations in households. Gender is used as an analytical tool to explore how different groups experience their environment on an everyday basis.

1 Assumpta Nnaggenda-Musana, "Housing Clusters for Densification Within an Upgrading Strategy. The Case of Kampala, Uganda" (PhD diss., Royal Institute of Technology (KTH), Stockholm, Sweden, 2008), 40.

2 Paul Isolo Mukwaya, "Urban Change in Uganda: The Challenge of Managing Urban Areas in a Global Environment," paper for the College of Urban Planning and Public Affairs (CUPPA) Conference at the University of Illinois, Chicago, USA, July 22–24, 2004.

3 Tom Sanya, *A Study of Informal Settlements in Kampala City* (Masters diss., Universität Stuttgart, Germany, 2001).

 | DOI 10.1163/9789004376793_007

The notable features of colonial urban planning in Uganda were economic and racial segregation. Racial segregation policies determined the growth of housing areas into British, Asian, and African neighborhoods in urban centers. Wielding power the British officials deemed themselves superior beings who came to modernize the indigenes ways of living. The Asian business and trades men were next down the rung and then the Africans, who mostly came to urban areas to do menial and other low paying jobs, were considered lowest in rung. Basing himself on Cape Town, South Africa—a country that was also colonized by the British—Graan observes that the discussions of architectural modernism most especially during the colonial era are 'strongly grounded in theories of power, power relations, mechanisms of control, and terrains of contestation and negotiation of difference.'[4] Graan further observes that the city council used segregation as a way of closely watching the living spaces of those that they controlled.[5] In Uganda, racially segregated housing was constructed in most of the district headquarters.

Housing for the British officers or *senior quarters* as they were called was located on hills[6] that allowed for good views, denoted power, and also ensured that the housing areas for the Asians or *junior quarters* and Africans—*African quarters*—could be easily monitored, or as in the South African context were 'clearly visible and controllable'.[7] Graan explains that in colonial South Africa, the modernist planning implemented by the colonial government *did not have* the concern for improving the wellbeing of its entire people, but rather became primarily a mechanism for social engineering, restructuring urban areas into racially and physically segregated areas, and dealing with issues around health and slum clearance, creating spaces that enabled the 'Colonial Other' to be noticeable and controlled.[8] According to Anderson and Rathbone, the houses for the British were situated on hills because they supposed that the high altitude would prevent the spread of air- and water-borne diseases.[9] In the South African context, Graan notes that categorizing people racially was used to draw out 'the different planning approaches that would be adopted by the City Council in the housing policies applied to different groups.'[10]

4 André van Graan, "Modernism as a Mechanism of Power and Control in Colonial Contexts: The Project of Modernity in Cape Town South Africa," in this volume.

5 Graan, "Modernism as a Mechanism of Power and Control," in this volume.

6 Sanya, *A Study of Informal Settlements in Kampala City*, 18.

7 Graan, "Modernism as a Mechanism of Power and Control," in this volume.

8 Ibid.

9 David M. Anderson and Richard Rathbone, *Africa's Urban Past* (Oxford: James Currey, 2000).

10 Graan, "Modernism as a Mechanism of Power and Control," in this volume.

In Kampala, the British housing areas dominated hills, such as Kololo, Nakasero, and Muyenga. Physical planning in the British neighborhoods was characterized by plots with ample space. These plots were measured in acres while in other parts of Kampala city, where the Asian and African households lived, the plots were measured in square meters. Today the plot frontages in the formerly British neighborhoods are still characterized by spacious plot subdivisions, ample public open spaces, and roads. Figure 6.1 shows the generous roads and road reserves that were designed for houses in the upscale Kololo hill commensurate to the first decade of the 20th century garden-city and modernist inspired planning ideals.[11] The houses on Kololo hill were isolated and sat on large gardens in the spacious plots, well provided with the basic infrastructure and services like roads, electricity, piped water, sewerage, and telephones lines.

The British neighborhoods were separated from the houses that were planned for the Asians and Africans by large open green areas bordered with wooded expanses or roads.[12] The Asian neighborhoods were further segregated from the African neighborhoods. The large open areas and wooded areas or roads mentioned above and the Asian neighborhoods buffered the British neighborhoods from the African settlements. The houses for the Asians including

FIGURE 6.1 *Kololo Hill, Kampala, Uganda, the generous roads and road reserves that were designed for houses in the upscale areas*
PHOTO: ASSUMPTA NNAGGENDA-MUSANA.

11 Nnaggenda-Musana, "Housing Clusters for Densification within an Upgrading Strategy," 27.

12 Sanya, *A Study of Informal Settlements in Kampala City*, 18.

the gardens and plots were smaller than those of the British but bigger than those for the Africans (see figures 6.2 and 6.3).

The different plots in the British neighborhoods were fenced off from each other with boundary walls demarcating the private territories of the

FIGURE 6.2 *Single storey house types in Asian neighborhoods, Kampala, Uganda*
PHOTO: ASSUMPTA NNAGGENDA-MUSANA.

FIGURE 6.3 *Double storey house types in Asian neighborhoods, Kampala, Uganda*
PHOTO: ASSUMPTA NNAGGENDA-MUSANA.

FIGURE 6.4 *African quarters in Naguru, Uganda, built in the 1950s*
PHOTO: ASSUMPTA NNAGGENDA-MUSANA.

FIGURE 6.5 *Distance between the African quarters in Naguru, Uganda, built in the 1950s*
PHOTO: ASSUMPTA NNAGGENDA-MUSANA.

British households. In the Asian and African neighborhoods, on the other hand, different households lived in semi-detached houses. The plots in the Asian neighborhoods had boundary markers like hedges to demarcate the territories of the different households, while in the African neighborhoods the outdoor space appeared more open and communal. The British intentionally designed the Asian and African houses in open spaces to be able to monitor and control these populations (see figures 6.4 and 6.5). According to Graan, 'the link between modern housing and mechanisms of control were clearly apparent.'[13]

Colonialism and Gendered Space

Like tents are to campers, it is argued that the African neighborhoods were small only because the African laborers—mostly men—were expected to live and commute from the rural areas. Majale notes that colonial government policy promoted the construction of houses with few rooms and with common facilities for the accommodation of short-term male laborers.[14] Graan states that in South Africa the Africans were provided with 'mainly hostel accommodation'[15] as far away as possible from family structures. The houses that were provided for them within the city were basically temporary lodgings. The room sizes were small with intentions of preventing the men from bringing their families to the urban areas.

Ironically many years later in the post-colonial period, urban planning in the major urban centers of Uganda is still guided by colonial planning principles and building regulations since no major modifications have been made to planning guidelines. Sengendo states that strategies to resolve housing problems in Uganda have not been comprehensive since they were founded during the colonial times.[16] As Mukwaya notes: 'Nothing much has changed from this pattern after the independence period.'[17] The government provided housing designs in Uganda—especially the informal settlements upgrading schemes—which intended to improve the shelter needs of the urban low-income households, seem to have been developed within the colonial urban planning framework.

13 Graan, "Modernism as a Mechanism of Power and Control," in this volume.

14 Michael Majale, "Settlement Upgrading in Kenya: The Case for Environmental Planning and Management Strategies" (PhD diss., University of Newcastle upon Tyne, UK, 1998).

15 Graan, "Modernism as a Mechanism of Power and Control," in this volume.

16 Hannington Sengendo, "The Growth of Low Quality Housing in Kampala–Uganda between 1972–1989" (PhD diss., University of Nottingham, UK, 1992), 65.

17 Mukwaya, "Urban Change in Uganda," 1.

The houses in such schemes have been referred to by some scholars as being smaller prototypes of colonial models.[18]

Like the situation was in apartheid South Africa, where the ways of living of the Africans were replaced 'with a "modern" vision of urban housing, that laid open and visible the threatening alleys and spaces that potentially fomented revolution',[19] the modernized house prototypes for low-income housing schemes in Uganda seem to have been designed as though these households need to be monitored. The forms and layouts of the housing prototypes 'make access, visibility, and observation easier'.[20] As a dire reminder of the African neighborhoods designed during the colonial times, the government provided houses are small as well as the plots, even if they are not intended to be occupied only by men like during the colonial times.

The sizes of the low-income houses are similar to those that housed men *temporarily* in the African neighborhoods during the colonial times regardless of new ideologies that encourage user-friendly and gender sensitive designs. These low-income houses have inevitably been retrofitted *permanently* with household sizes that were not considered in the original designs. In other words, these households, which include women, children, and extended family, have had to squeeze into spaces that were not tailored or designed for them. It is interesting to see how the women, children, and extended family have adapted to these houses.

Designing for a particular sex, race, class, or ethnic group brings a gender element to design. The term gender, in many cases, has been associated with the biological differences between men and women making it synonymous to the term sex. Larsson defines gender as a term that 'denotes what a society describes to be male and female, varies over time and place, a social and cultural construction.'[21] Gender can as well refer to the economic, social, and political divergences involving women and men, including their perceptions of the quality of life and living opportunities.[22] Being socio-culturally contextualized, Mapetla and Schlyter emphasize race and class as important variables in gender relations.[23] Gender concerns among the women who belong to a particular

18 Sengendo, "The Growth of Low Quality Housing in Kampala–Uganda between 1972–1989."

19 Graan, "Modernism as a Mechanism of Power and Control," in this volume.

20 Ibid.

21 Anita Larsson, "Gender Perspectives in Housing and Planning," *Building Issues* 11, no. 1 (2001): 1–18.

22 Nnaggenda-Musana, "Housing Clusters for Densification within an Upgrading Strategy," 96.

23 Matšeliso Mapetla and Ann Schlyter, *Changing Gender Relations in Southern Africa. Issues of Urban Life* (Roma, Lesotho: Institute of Southern African Studies, National University of Lesotho, 1998), 3.

group might be different from those of another group. Larsson draws attention to the close relationship between the concept of gender and physical planning and housing design. She notes that the gender concept 'can be used to create a woman-friendly environment, and can be used to analyze women's subordination in society.'[24]

In the context of the MWHS: Masese Women's Housing Scheme, the houses that were designed to accommodate men and are now housing women clearly reflect gender biasness against women and stand as a symbolic evidence of how space can be *gendered*. Gendered space hereby refers to space that is expected to be used by one sex or otherwise accommodates activities that are assumed to be performed by different sex groups. Spain argues that gendered spaces can provide a solid, everyday life foundation for the status difference to be produced and transformed within different socio-cultural contexts.[25] With that understanding, it could be argued that the housing designs and the spaces that were produced under the colonial planning principles and regulations are gendered. These houses can be seen to be gendered in relation to race as we have noted above as well as in relation to sex as is the case in MWHS. The MWHS houses are evocative of the houses that were provided for the African male laborers during the colonial period.

Women Living in Male Space

Space can be considered as an ordering principle and so is gender. Space represents social organization. Space speaks; houses have been associated with women and have been the loci of female labor and the daily tasks of food preparation, childrearing, and maintenance of the physical and cultural household. According to Spain, parochial space extends the porosity of public and private—and their associated gender roles—beyond the boundaries of a discrete household to the communities in which they were situated.[26] The social construction of gender differences creates women and men spaces. Domestic space is termed as private space, thus, the sphere in which women operate.[27] Women are seen as being concerned with the particular.[28] In MWHS the

24 Larsson, "Gender Perspectives in Housing and Planning," 5.

25 Daphne Spain, *Gendered Spaces* (Chapel Hill, NC: University of North Carolina Press, 1992), 233.

26 Spain, *Gendered Spaces*, 7.

27 Ibid.

28 June Jordaan, "Architectural Agency and 'Place-Making' in a Transformative Post-Apartheid South African Landscape," in this volume.

gender roles in domestic space were altered. The boundaries between spaces are blurred, there is no clear division 'between private, parochial, and public spaces'.[29]

At the Neighborhood Level

The MWHS: Masese Women's Housing Scheme in Jinja is one of the government provided informal settlements upgrading schemes in Uganda during the late eighties. The Ugandan government was facilitated by organizations, such as the Danish International Development Agency (DANIDA), United Nations Development Program (UNDP), United Nations Centre for Human Settlements (UNCHS), and Shelter Afrique to develop these informal settlements upgrading schemes. These schemes include Namuwongo in Kampala, Malukhu in Mbale, Oli in Arua, and MWHS in Jinja.[30] This study focuses on the MWHS because unlike the other informal settlements upgrading projects, the MWHS is unique in that it targeted women as the main beneficiaries. This project was provided mainly for low-income women to avail them with modern housing probably as a relief from the inadequate shelter in the informal settlements where they had been living. The MWHS was conceived by the Ugandan government as confirmation of its commitment towards mainstreaming gender in housing. This followed the Third World Conference on Women (TWCW) that was held in Nairobi, Kenya, in 1985, whereby the East African countries dedicated themselves to promoting gender issues in all government sectors, including housing. Seemingly the TWCW impacted positively on raising gender awareness in Uganda through the formation of MWHS. However, were the houses adequate? Did the planners and architects of the MWHS house prototypes stop to think about the occupants of these spaces?

According to Jordaan, 'the role of the architect is to provide good "form" wherein communities can be stabilized, dwelling can take place, and thereby produce happy, non-alienated, non-conflicted human beings.'[31] Comparing the MWHS house prototypes to the colonial housing designs for the Africans it seems that neither the planners nor the architects of MWHS thought about the production of 'non-conflicted'[32] places. As if they were to allow for adequate surveillance, the plots of MWHS were designed in a gridiron kind of layout in small sizes of 18m × 12m. These plots were arranged around an open space facing a public road in clusters of 8–18 plots (see figures 6.6).

29 Spain, *Gendered Spaces*, 7.

30 Nnaggenda-Musana, "Housing Clusters for Densification Within an Upgrading Strategy," 58–59.

31 Jordaan, "Architectural Agency and 'Place-Making'," in this volume.

32 Ibid.

FIGURE 6.6 *Model of a cluster housing at the MWHS: Masese Women's Housing Scheme*
AUTHOR: EIMAN AHMED ELWIDAA

The house prototypes for the MWHS neighborhood were meant to be enclosed by boundary walls around the plots to demarcate the extent of each plot. However, today these walls have not been built. Did the women in the MWHS not construct the boundary walls due to lack of money to build them? The women demarcate their different plots with short hedges to mark their territories. They create their own boundaries albeit differently. Jordaan notes that a boundary 'is that point where something begins its "presencing"'.[33] Some women state during interviews that even if they were able to afford building the boundary walls, they would not, because they feel that these walls can contribute to social alienation. To the MWHS women, a neighborhood that enables them to interact with each other is an essential component to the convenience of that setting. In the MWHS context, the neighborhood concept should be understood to mean any household within five to ten minutes of walking distance. In some circumstance, the households are not only neighbors, but are also relatives or behave like relatives even if they are not. The women feel that the walls are a barrier to good social interaction, which is essential for their wellbeing and satisfaction. In order to sustain the social interaction households visit each other frequently. Sometimes the visits are formal or informal. Informal visits can be in the form of borrowing salt for cooking. Reciprocal visits are part of social interaction with people from different households. Formal and informal visits influence the continuation of households bonds, and enable the formation of new fields of interaction.

33 Jordaan, "Architectural Agency and 'Place-Making'," in this volume.

When discussing architecture and gender, Spain argues that residential architecture, among other landscape features, 'reflects ideals and realities about relationships between men and women within the family and society.'[34] In MWHS, absence of boundaries between houses, coupled with the housing transformations, has resulted in the emergence of organic spatial patterns that are in departure from the assumed ones. The emerging organic spatial patterns only hint at the original planning layout but defeat the prior rigid gridiron arrangements that were stipulated at the onset of the housing scheme. There is no clear demarcation of private and public space, since private space seems to spill over into the surrounding public spaces. The women's rejection of the construction of boundary walls can be viewed as repulsion against fixation to confinement and a pursuit for liberation and freedom. Women extend their realm beyond the captivity of their homes, and extend themselves beyond boundaries to new horizons and in a solid statement that denies submission and control. The houses are small in size and congested spaces inside, which forces the women to receive their visitors outdoors. Therefore, the outdoor space of the house becomes a reception area to compensate for the small space within the house. With the absence of demarcating fences between these houses, social bonding is maintained.

One might wonder what the rationale behind the phenomenon of boundaryless living is. Is it just a representation of women emancipation or is there more to it? What can we learn from this phenomenon that can assist architects and planners to design 'happy, non-alienated, non-conflicted'[35] housing that can enhance people's wellbeing? It is observed that the newly emerged physical setting of MWHS—whether conscious or subconscious—is conducive to women's housing needs and their ways of living. The ideals of true womanhood—as domesticity was alternatively known—elaborated women's position within the private sphere and celebrated qualities, such as piety, purity, submissiveness, and domesticity.[36] This ideology segregated male and female responsibilities. Women of MWHS feel more secure in their neighborhood because the absence of boundary walls creates defensible space allowing for natural surveillance among neighbors, which makes them feel protected. Women can embrace power within their own realm even though modernization seems to marginalize them even more. The MWHS women observed that absence of barriers and restrictions gives them a sense of belonging and social existence that is comforting and makes them *happy* human beings.

---

34 Spain, *Gendered Spaces*, 7.

35 Jordaan, "Architectural Agency and 'Place-Making'," in this volume.

36 Janet Giele, *Two Paths to Equality: Temperance, Suffrage, and the Origins of Modern Feminism* (New York: Twayne, 1995), 36.

At the House Level

In MWHS, there were three types of plots, which came in sizes of small, medium, and large. Three different house prototypes were designed on each type of plot. The women, who benefitted from this scheme, were supposed to choose their house designs from government provided prototypes and based on their capacity to pay back the loans that were provided to cover the house construction expenses. The project stipulated that the medium and large plots could be accessed only by those women, who had formal employment in the formal sector. However, most women who were to benefit from this scheme were working in the informal sector and could only afford the small houses on the small plots. Thus, the small houses, which are frequent in the MWHS neighborhood, are the focus of this study.

The small houses prototypes were designed in three types 1, 2, and 3 (see figure 6.7). The layout of the houses comprised of two rooms—one to act as a living room and the other as a bedroom—and a corridor as in type 1. In type 2, part of the corridor was reduced and a bathroom was introduced. In type 3, one of the rooms was enlarged by extending it into the corridor area. The remaining space between the enlarged room and the bathroom then formed a shade. However, these spaces have been transformed to date (see figure 6.8 where the house on the left has been extended on the extreme right).

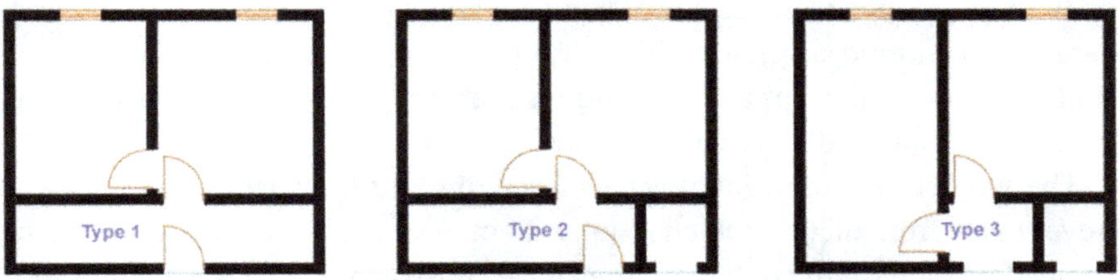

FIGURE 6.7 *Plans of the MWHS house prototypes*
AUTHOR: EIMAN AHMED ELWIDAA.

FIGURE 6.8
MWHS house that has been extended on right side, in Jinja, Uganda
PHOTO: EIMAN AHMED ELWIDAA.

Space in the MWHS neighborhood can be divided in four types: private, semi-private, semi-public, and public. Women and men can be observed in these four space types. The private space is that which is immediate to the houses, this includes the indoor and outdoor space. The semi-private spaces are those, which people close to the household can access. The semi-public space is that space closer to the houses—like in cases where Home Based Enterprises (HBEs) are present—and is accessed by complete strangers who come to buy groceries. Public space is accessed by strangers. This is unlike in the traditional modern housing where private and semi-private space is found within the plot, and the private spaces like kitchens are defined as feminine and are arenas for women's work, including female members of the household and the house helps. The kitchens in this case are usually found in the private spaces behind the houses. According to Ardener, the 'notion of "private" as opposed to "public" is seen as a criterion for "mapping" metaphysical space, as "inner" does in opposition to "outer".'[37]

In the MWHS houses,—unlike traditionally when women were relegated to the backyard—the women are often found in the front yard. Thus, the MWHS private and semi-private space can be found at the front of houses. The front of the houses and the plots are multifunctional areas, and are women's spheres of influence. This is where these women interact with neighbors or with customers in scenarios where HBEs are located in front of the houses. Women also do their chores like washing clothes and cooking food in front of their houses, here they can tend to their children, do beauty care—like hair plaiting—or chat. Some can take naps too while others are working. Dishes are washed outdoors in a traditional way in a basin or bucket.

The women use traditional ways of cooking, for instance, using charcoal stoves made from oil tins, which require them to sit while cooking. Some women need privacy when cooking so they have constructed shades. Luckily, these shades are open to draft preventing heating up and allowing for a cool atmosphere in the cooking shades and enabling the women from different households to do their house chores while chatting with neighbors.

The ambitions of most women are to acquire enough space. Since most of them earn meager wages, their wishes are to be able to extend their houses so as to acquire income generation space, to be able to earn sufficient extra money to care for their households. Transformations can include rooms for rental as accommodation, salons, eating spaces, or bars. The alterations are also made to create more room for user households, for example, rooms for

37 Shirley Ardener, *Women and Space, Ground Rules and Social Maps* (Oxford: James Currey, 2000), 4.

extended family. Associations with close kin are important to MWHS households combining economic security, political influence, social support, and psychological assistance in the form of emotional security and in particular entertainment. An effective field on interaction can occur between neighbors of the same social status, and friends, these become kin by affinity.

Two main forms of transformations have been noted in MWHS. The first was mainly in the form of adjusting the internal space by zoning off new space within a larger space to accommodate variant activities of the households within the same space. Women, prompted by their traditional gender role as a homemakers, used the internal spaces creatively by for instance, extending the limited indoors space to the roof space to act as a storage area for their home accessories and belongings.

Transformations have been made on the outside of the houses in the form of semi-attached or detached extensions to accommodate rooms. These rooms are usually built to house extended family or older children—especially boys—who are expected to be separated from their younger siblings in the African culture. The internal space of the houses has been sub-divided in some cases to accommodate tenants who can pay the households some money monthly. In figure 6.9 a woman, who earns extra income from selling jerry-cans of water uses the roof area to store her merchandise on top of the trusses.

In MWHS, traditional, patriarchal, extended family still prevails. However, matrilineal tendencies cannot be disregarded. Women gain status, prestige, and relative freedom when they are self-sustaining financially. The low-income women of MWHS, beside their conventional gender role of taking care of the families by being responsible for cooking, child rearing, and other domestic chores, indulge in HBE to increase their income. In female headed-households, women are the sole income earners in the households forcing them to take both roles of the caretaker and the breadwinner. To be able to meet all of those

FIGURE 6.9
Roof storage space
PHOTO: EIMAN AHMED ELWIDAA.

obligations, necessitate women to develop their multi-tasking skills. It is a common scenario in MWHS for women to attend to customers from a distance while washing clothes or watching over their cooking food. The goods often sold are handmade crafts or groceries grown in gardens around their houses.

It is important to note that most of the women's activities are performed outdoors. The indoor areas are used in case of adverse weather conditions. The intense use of outdoor space could have been dictated by the limited sizes of the indoor space and the way the activities are conducted. Cooking on charcoal stoves, washing and drying clothes, and attending to shops, kiosks, neighbors or visitors, are all done outdoors. The absence of boundary walls between plots has had a positive impact on the women as it enables them to carry out their chores within the vicinity of their houses while chatting with each other.

The two-roomed houses of MWHS were not sufficient to accommodate the households. It is important to state that in Uganda and in the MWHS sociocultural context the phenomenon of house being able to receive extended family is familiar and essential. However, the designs of the house prototypes of MWHS seem to be based on colonial perceptions expecting the house to be occupied by nuclear family households. These houses are not designed to include extended, polygamous, or female-headed households, commonly found in the African cultures. Evidencing that housing should be a reflection of its users' values, aspiration, and motivations, MWHS women had to transform their housing space and adjust it to accommodate their housing activities and needs. Since most of them earn meager wages, their wishes were to use their houses in a way that could promote their welfare, for instance, by helping them increase their income to care for their households. This confirms Larsson's argument that the term *housing* has a broader meaning, embracing both its value as a shelter as well as its social and physical surrounding, and as such housing becomes crucial for women as a center for their activities.[38]

Updating Low-Income Housing

Space without boundaries influences the kind of social interaction that is seen in MWHS. Social identity in MWHS is determined by the physical and spatial constituents of space. In MWHS, it is important to note, however, that an absolute separation of spheres does not exist. Space is not used as it has been socially constructed, whereby the public space is occupied by men and the private by women. Gender-specific as well as sexually integrated space extends

38 Larsson, "Gender Perspectives in Housing and Planning," 4.

beyond the walls of domestic space into the outdoor areas. Blurred space or 'spaces within spaces or overlapping universes'[39] is a common phenomenon in MWHS. These social functions illustrate that the use of space in MWHS is fluid and contingent upon the type of social interaction and the human agents present in that space. In low-income housing, women have been noted to be physically dominant in domestic space that was originally masculine.

Evident in MWHS are adaptations that have developed over time, these are alterations in government provided house prototypes and the new transformations. The women in most cases are continuously transforming the houses that were provided for them. Kishindo notes that like in several African cultures in which women are assigned the role of turning a house into a home, the MWHS women managed to transform the impersonalized houses to make them a *home* that promotes their existence and bloom.[40]

The houses in the MWHS neighborhood are small and appear to be similar to the African neighborhoods. However, most of the houses have been transformed. Why are the houses being transformed? Is it because they were designed for male habitation? How can these transformations inform us? Were the ways of living of the MWHS women considered when designing their housing? Jordaan argues that space should be bound with reality. Places should be designed with relevance to reality.[41] The appreciation of physical space is dependent on the social perceptions of it. How are roles of men and women shaped in housing space? Ardener explains that 'objects are thought to structure the environment immediately around them.'[42] However, it can be observed that the MWHS houses took on a physical form not determined by the women even though they are meant to live in them perpetually. Most houses in the MWHS have been transformed from their original forms. The transformations include extensions to houses to accommodate various activities. Most of these activities are related to income generation. The MWHS was meant to house women that had been liberated from the harsh conditions in the informal settlements; in a way, the scheme should have been designed as a *healing instrument*.[43] However, the housing transformations seem to tell a different

39 Ardener, *Women and Space, Ground Rules and Social Map*, 3.

40 Emma C. Kishindo, "Women and Housing Insecurity in Malawi: A Study of Two Low-Cost Housing Projects in Zomba District in Malawi," in *Gender and Urban Housing in Southern Africa: Emerging Issues*, eds. Anita Larsson, Matšeliso Mapetla, and Ann Schlyter (Roma, Lesotho: University of Lesotho, 2003).

41 Jordaan, "Architectural Agency and 'Place-Making'," in this volume.

42 Ardener, *Women and Space, Ground Rules and Social Maps*, 3.

43 Jordaan, "Architectural Agency and 'Place-Making'," in this volume.

story that probably the Masese women might have been dissatisfied with the houses.

Interesting to note is that the house designs, the building materials, and construction techniques are in no way similar to those that were intended for the project. The construction materials and techniques for most of the transformations are indigenous, for example, adobe, and mud and wattle. It is as though—through the assertion of their socio-cultural identity—the women of MWHS are showing off their dignity and pride. These women are not 'weak' or 'emasculated'.[44] By transforming their houses, they are in fact resisting the original small houses that were imposed on them. The universal and universalizing space of modernism can be regarded as a culprit, which generates disaffection in society.[45] The house transformations in MWHS may indicate that the prototypes that were provided by government are inappropriate. Government tried to capture a state of perfection different from that in the informal settlements from which the women came, by creating an environment that was planned in a gridiron manner. The transformations that are developing in the MWHS are organic in nature exemplifying 'an ability to change and adapt to individual needs'.[46]

This chapter emphasizes the need for design professionals—planners and architects—to indulge their clients in the design process of their houses. Jordaan notes that the role of architects should be to provide 'good "form" wherein communities can be stabilized' instead of reproducing western ideals.[47] How can better designs for low-income households be developed today to evidence a departure from colonial planning and design? Seeing that the government provided housing schemes have been deemed a failure[48] since obviously the beneficiaries are not satisfied, the Ugandan planners, architects, and other building professionals should look back into colonial planning and try to 'track the failures, silences, displacements and transformations produced by its functioning.'[49] Housing has never been static as it changes over time and through transforming conditions. Designing houses for low-income

44 Graan, "Modernism as a Mechanism of Power and Control," in this volume.

45 Jordaan, "Architectural Agency and 'Place-Making'," in this volume.

46 Ibid.

47 Ibid.

48 Nnaggenda-Musana, "Housing Clusters for Densification within an Upgrading Strategy," 15.

49 Garth A. Myers, "Intellectual of Empire: Eric Dutton and Hegemony in British Africa," *Annals of the Association of American Geographers* 88, no. 1 (1998): 1–27 sited in Graan, "Modernism as a Mechanism of Power and Control," in this volume.

households calls for inspirations, aspirations, and ideologies that promote, celebrate, and advance their wellbeing and prosperity.

Currently MWHS has developed in contrast to how it had been envisaged to grow. Like the beauty one finds in handmade products over machine produced ones, the women of MWHS have manipulated their neighborhood to tame and develop it according to their ways of living. They have, thus, turned MWHS into a vivid, self-revealing, and emancipating living environment that celebrates their life, class, and race to take it beyond the borders of reticence, confinement, and suppression. Physical settings that facilitate associations with close kin, neighbors, and each other are a significant *reservoir* to low-income households.

Bibliography

Anderson, David M., and Richard Rathbone. *Africa's Urban Past*. Oxford: James Currey, 2000.

Ardener, Shirley. *Women and Space, Ground Rules and Social Maps*. Oxford: James Currey, 2000.

Byerley, Andrew. *Becoming Jinja. The Production of Space and Making of Place in an African Industrial Town*. Stockholm: Almqvist & Wiksell International, 2005.

Byerley, Andrew. *Displacements in the name of (Re)Development: The Contested Rise and Contested Demise of Colonial 'African' Housing Estates in Kampala and Jinja*. Stockholm: Intellecta DocuSys AB, 2005.

Byerley, Andrew. "Mind the Gap! Seeking Stability Beyond the Tribal Threshold in Late-Colonial Uganda. The Role of Urban Housing Policy 1945–1960." *African Studies* 68, no. 3 (2009): 429–464.

Byerley, Andrew. "Ambivalent Inheritance: Jinja Town in Search of a Postcolonial Refrain." *Journal of Eastern African Studies* 5, no. 3 (2011): 482–504.

Giele, Janet. *Two Paths to Equality: Temperance, Suffrage, and the Origins of Modern Feminism*. New York: Twayne, 1995.

Goodfellow, Tom. "'The Bastard Child or Nobody?': Anti-Planning and the Institutional Crisis in Contemporary Kampala." *Crisis States Research Centre Working Papers Series* 2, no. 67 (2010): 1–28.

Kamete, Amin Y. "Interrogating Planning's Power in an African City: Time for Reorientation?" *Planning Theory* 11, no. 1 (2011): 66–88.

King, D. Antony. *The Bungalow. The Production of a Global Culture*. London: Routledge and Kegan Paul, 1984.

Kishindo, C. Emma. "Women and Housing Insecurity in Malawi: A Study of Two Low-Cost Housing Projects in Zomba District in Malawi." In *Gender and Urban Housing in Southern Africa: Emerging Issues*, edited by Anita Larsson, Matšeliso Mapetla, and Ann Schlyter, 74–99. Roma, Lesotho: Institute of Southern African Studies, National University of Lesotho, 2003.

Larsson, Anita. "Gender Perspectives in Housing and Planning." *Building Issues* 11, no. 1 (2001): 1–16.

Larsson, Anita. "From Equal Opportunities to Gender Awareness in Strategic Planning. Reflections based on Swedish Experiences." *Town Planning Review* 77 no. 5 (2007): 507–530.

Majale, Michael. "Settlement Upgrading in Kenya: The Case for Environmental Planning and Management Strategies." PhD Dissertation, University of Newcastle upon Tyne, UK, 1998.

Mapetla, Matšeliso, Anita Larsson, and Ann Schlyter. *Changing Gender Relations in Southern Africa. Issues of Urban Life*. Roma, Lesotho: Institute of Southern African Studies, National University of Lesotho, 1998.

Mukwaya, Isolo Paul. "Urban Change in Uganda: The Challenge of Managing Urban Areas in a Global Environment." Paper for the College of Urban Planning and Public Affairs (CUPPA) Conference at the University of Illinois, Chicago, USA, July 22–24, 2004.

Nnaggenda-Musana, Assumpta. "Housing Clusters for Densification Within an Upgrading Strategy, The Case of Kampala, Uganda." PhD Dissertation, Royal Institute of Technology (KTH), Stockholm, Sweden, 2008.

Omolo-Okalebo, Fredrick, Tigran Haas, Inga Britt Werner, and Hannington Sengendo. "Planning of Kampala City 1903–1962: The Planning Ideas, Values, and their Physical Expression." *Journal of Planning History* 9, no. 3 (2010): 151–169.

Sanya, Tom. "A Study of Informal Settlements in Kampala City." Masters Dissertation, Universität Stuttgart, Germany, 2001.

Saunders, Kriemild. *Feminist Post Development Thought. Rethinking Modernity, Post-Colonialism and Representation*. London: Zed Books.

Schlyter, Ann. *A Place to Live. Gender Research on Housing in Africa*. Uppsala: Nordiska Afrikainstitutet, 1996.

Sengendo, Hannington. "The Growth of Low Quality Housing in Kampala–Uganda between 1972–1989." PhD. Dissertation, University of Nottingham, UK, 1992.

Simone, AbdouMaliq. "Spectral Selves: Practices in the Making of African Cities." Paper presented at the Wits Institute for Social and Economic Research and New School University, Johannesburg, South Africa, 2002.

Spain, Daphne. *Gendered Spaces*. Chapel Hill, NC: University of North Carolina Press, 1992.

PART 3

Spirituality and Decay in Architecture

CHAPTER 7

The Heterotopic Nature of the Built Heritage: The Sacred Wooden Architecture of Transylvania and Its Practices

Smaranda Spânu

The Built Heritage Object as a Hierarchical Space

The values attached to the built heritage—be they historical, esthetical, social, economic, etc.—are usually accumulated in time through an externally imposed process.[1] Yet these values are not rigid or immutable hierarchies and, as cultural history and preservation history reveal, they are constantly shifted and rearranged according to the specific context within which the built object is being perceived. This chapter argues that these hierarchies of value are not simply erased and replaced as the built object changes its context; they are juxtaposed and often remain imprinted within the physical form of the object, a process that is most visible in the case of immovable heritage rather than the movable one. Such objects cumulate their cultural meanings just as additions, transformations, extensions, reconstructions, etc. are adding to the overall image of the object.

The overall value of the object is a cumulative process that can be roughly divided into three phases: a first layer of meaning, generally defining the object's architecture and overall physical form, derives from its original function and original patterns of practice. This layer of meaning is inherent to the building process. Depending on the functional category of the object, this initial layer of meaning can be associated to a representational value, intrinsic to the object, and different from a value gained in time. As a general rule, this layer of meaning is the most resilient one, as well as the most important one, in terms of documentary value. The representational value is common, as well as best preserved, in the case of some particular groups of built heritage such as aristocratic residencies and institutional or religious buildings. In the second phase,

1 Lisanne Gibson and John Pendlebury, "Introduction: Valuing Historic Environments," in *Valuing Historic Environments,* eds. Lisanne Gibson and John Pendlebury (Surrey: Ashgate Publishing, 2012), 1; quoting Riegl's structure of values, Kazemer Kovacs, *Timpul monumentului istoric—The Time of the Historical Monument* (București: Paideia, 2003), 15.

 | DOI 10.1163/9789004376793_008

meaning is assigned through a process which generally implies temporal accumulation: the community establishes a specific practice pattern for a specific object, thus endowing it with additional cultural, social or historical signification. The practice of a particular object (i.e. one of representational value) can unfold as originally *prescribed* by its initial function, apparently conserving its initial meaning. The process of temporal accumulation is in this case a subtler one: the meaning initially conferred is passed on through continuous reassertion within different contexts—whether by different generations of the same community or between different communities. This process implies variable attitudes towards the built object, most of which contribute to the physical alteration of the built object, whether by adaptive interventions (such as enlargements, extensions, additions, restorations, etc.) or by continuous everyday practice (what is usually referred to as functional wear).

As the social, cultural, historical, economical, and technological context changes, 'the dominant frameworks of value of the time and place'[2] change as well, and along with them the status of the built object. As it is passing through this process, the object accumulates some layers of meaning while it seems to be losing others; yet these *disappearing layers* simply lose their visibility as they are pushed in the background. Most often the shifting of these layers has a direct impact on the material form of the object, in some cases even leading to the objects' complete extinction.

The hierarchies of value attached to the object are the basis of its identity and its status, thus becoming subject to change. When values are redefined, the wider notion of heritage is redefined as well. The modern contemporary heritage policies can be seen as a lens, through which the built object is being acknowledged with all of its cumulated significations, both central and secondary ones, as well as its complex constructed value. While a hierarchy is necessary in order to manage the multiple manifestations of built heritage, the modern preservationist perspective tends to and aims to acknowledge all significations ingrained at one time or another in the fabric of the object. This tendency can easily be classified as post-modern, as it aims at reconciling multiple and frequently antagonistic readings that reside in the built object; different cultures and actors, different time intervals and events as well as the quite prevalent double-coding,[3] all add to the divergent nature of heritage

2 Gibson and Pendelbury, "Introduction," 1.

3 The notion of double-coding is used here in order to describe the specific situation in which a certain heritage object is endowed with both positive and negative meanings, by opposing actors, thus becoming the subject of a simultaneous double reading; intrinsic to this condition the specific object is constantly negotiated, challenged, and its physical form threatened.

significance, yet the protected status is projected as an all-enveloping, overarching significance, confirming and instituting all of the inscribed meanings. The history of preservation demonstrates the evolution of this perspective through the extended and still continuing process of defining and enlarging its areas of interest. The search for an overarching notion seems to find its resolve only at the end of the 20th century, when the concept of 'cultural significance'[4] was coined in the 1999 Burra Charter by Australia ICOMOS, in order to define (as an integrated notion) the numerous values accumulated by the heritage object. On a closer inspection one can situate this protectionist construct within a broader context, revealing it to be just another way of reading, constructing, and interpreting the objects values, creating hierarchies—in order to assess the status of an object or of a class of objects. They themselves can be seen as 'just another layer of meaning'. Since the protected status is conferred to the object by the community (be it local or global community), and since it operates on the physical form of the object as well as on its perception within the community, it joins all of the other previously accumulated layers.

The contemporary modern protectionist interpretation deems itself the final layer in an attempt to hinder the destructive-evolution of the object; these policies are defying the natural and inevitable course of time, and yet they acknowledge them as immutable. Despite being aware of their paradoxical condition, we are continuously redefining their paradigm.

As Delafons observes, conservation as well as preservation and restoration—with their policies and practices—have acquired quite an extensive history, and their scope 'has widened enormously over the past hundred years and diverse influences have shaped [their] development'.[5] Very diverse classes of heritage engage a myriad of layers of meaning, actors, and multiple-encodings. The accelerated process of 'heritage-ization'[6] or 'monumentification'[7] that we

The actors involved are multiple, varying from opposing ethnic communities to subgroups or single individuals within a community.

4 The Burra Charter (Australia ICOMOS, 1999).

5 John Delafons, *Politics and Preservation—A Policy History of the Built Heritage, 1882–1996* (London: E & FN Spon, an imprint of Chapman & Hall, 2005), 3.

6 Heritage-ization or heritageification from *'patrimonialisation'*, fr.—defines the 'process through which elements of culture and nature become, at some point in the history of societies invested with the quality of heritage worthy of being saved, showcased for the benefit of present generations and passed on to future generations'. Ahmed Skounti, "De la patrimonialisation. Comment et quand les choses deviennent-elles des patrimoines?", *Hesperis-Tamuda* vol. XLV (2010): 19.

7 *Monumentification*—the term is defined by Kazemer Kovacs as a 'tendency to confer the memorial value to almost anything created by human civilization in the past' or the tendency

are currently witnessing reflects the globalization of particular patrimonial mentalities, considered by some as a modern symptom of the identity crisis of the human constructing capacities.[8] It is also a symptom of the assumed post-modern plurality. By acknowledging multiple hierarchies of value, the protectionist policies acknowledge and validate multiple identities. Since policies are constructed, generally speaking, to regulate the nonconforming, to counter a specific undesired, negative course of action, the nature of these protectionist policies paints a gloomy picture of the contemporary state of heritage. These policies are born from the will to challenge the all-levelling globalizing tendencies and the overpowering market forces that operate within the contemporary society, and that particularly afflict built heritage. Yet, despite their motives being beyond doubt, their functioning is; some situations stand witness to the hijacking of preservation policies in service of economical mechanisms.

This chapter argues that the 'heritage-ization' phenomenon reflects a deeper cultural identity crisis. We are faced with the general disappearance of heritage (mobile, immobile, and immaterial) in the form of the rapidly thinning of urban protected fabric, the deterioration of cultural landscapes, or the vanishing of entire protected sites, collections, and categories of objects. The counter reaction manifests in the form of forestalling further degradation and salvaging as many material cultural identity markers as possible, even if their practices, cultural schemas, and contexts no longer exist.

The built object is simultaneously a product of cultural practices and of certain politics and power-relations—that construct a specific space and its contents, in a specific period. As these shift, such culturally-loaded objects become heritage-protected items. The protection targets the conservation of imprints of most often disappeared or threatened practices onto a material form—this being in itself a practice of experiencing the space. The heritage object becomes an isolated enclave, removed from the natural flow of time, thus becoming other in relation to its context. Such is the case of the historical wooden churches of Transylvania, Romania. These churches have become hybrid spaces—both sacred ritual spaces and free-access desacralized public-spaces—and mirrored spaces: in most cases a contemporary church is flanking the old one. As a product of historical politics and power-relations, these churches were forcibly built to disappear, thus their perishable nature. This politics of invisibility shifted to visibility as their identitary-patrimonial status was recognized. They have become contested spaces—being denied

to 'identify a historical monument in almost every artefact whose fabrication technology has become obsolete'. Kovacs, *Timpul monumentului istoric*, 128–129.

8 Kovacs, *Timpul monumentului istoric*, 116.

their original function: in order to preserve the imprint of cultural practices, these practices are prohibited within the protected space. Although sharing the practices that generated them, the protectionist contemporary policies and the current practices within the communities to which they belong, these wooden churches evolved differently. This chapter focuses on a comparative analysis of such cases in an attempt to identify the relationship that led to such radical evolutions.

The 'Heterotopic' Profile

The succinct sketching of the concept of heterotopia provided by Foucault in one of his short unpublished works focused the attention of philosophers, sociologists, anthropologists, psychologists, geographers, and finally architects, despite often being accused of ambiguity. Its apparent looseness has spawned numerous readings and interpretations. This distinctive type of space, described by Foucault through an imported medical term, was further defined through historical examples illustrating its multiple and distinct manifestations and its alternate orderings and ways of functioning. These places constituted their functional and spatial otherness through the relation to their context, and as Hetherington argues 'their presence either provides an unsettling of spatial and social relations or an alternative representation of spatial and social relations'.[9] He then identifies three main yet overlapping stances of the meaning of *other*: 'something without (defined as different to the norm either within a culture or between cultures…), something excessive or something incongruous, a hybrid combination of the incongruous';[10] besides these one can always encounter different manifestations of the otherness. In Hetherington's interpretation, heterotopic space isn't reduced to physical space or to virtual space; the heterotopic profile is constructed through the juxtaposition of the specific materiality of a place, of social practices and particular events. Heterotopic character is thus simultaneously place-bound, socially-bound, and time-bound; if one of these 'compounds' is missing and one of these relations is disrupted, the heterotopic character can no longer exist.

> It is the heterogeneous combination of the *materiality*, *social practices* and *events* that were located at this site [i.e. Palais Royal] and what they

9 Kevin Hetherington, *The Badlands of Modernity: Heterotopia and Social Ordering* (London and New York: Routledge, 2002), 8.

10 Id., 8.

> came to represent in contrast with other sites that allow us to call it a heterotopia. Heterotopia exists when the relationship between sites is described by a difference of representation defined by their modes of social ordering.[11]

I argue that Foucault's concept as well as Hetherington's interpretation of it can be introduced and usefully applied to the built heritage space. While Hetherington dismisses the importance of the individual input of physical space to the heterotopic character, Foucault's descriptions contour specific palpable characteristics that can be read through an architectural lens. I argue that a built object can harbor a heterotopic potential within its material and spatial characteristics, since these are the direct outcome of specific social orderings. A built object is created with a specific ordering in mind, and thus it becomes the physical representation of that ordering; if this ordering *is other* or *becomes other*—opposed, different, or incongruous in relation to its context—it triggers the heterotopic character of that object. Yet, the alternative ordering that generates a built object remains embodied in that specific form even after the practices and events that completed its heterotopic character have gone. As the object cumulates layers of meaning, as well as other orderings, its potential heterotopic character can become reactivated. When translating this in the domain of built heritage one can identify two instances of heritage as other: (a) the object, heterotopic through its original function, form, and practices (or through its accumulated ones)—is acknowledged as valuable and treated as such; its alternate ordering, its *otherness*, is the main rationale behind its subsequent classification; (b) the concept of heritage, as defined by the global and local community, is introduced as an alternative (heterotopic) ordering within specific spaces that are considered valuable, with the deliberate purpose of conserving their meanings; the built object becomes heterotopic through its newly acquired protected status, even though originally it didn't have an alternate ordering. In the first instance the otherness of the object usually entails a certain degree of protection, its community organically and intuitively ensuring its survival—thus instituting a bottom-up preservation, as well as heterotopic character. In the second instance, the official status is top-down introduced. When these two overlap the heterotopic character is doubled and all its specific characteristics are emphasized.

This first instance appears to be more suitable for a case by case analysis or a comparative one. The second instance, which will be investigated further on, can provide a more abstract tool for analysis, allowing the identification

11 Ibid.

of prevalent, specific heterotopic features of the built heritage. These features considered as an ensemble construct the heterotopic profile of built heritage.

The Heterotopic Characteristics of Built Heritage

During these last decades the contemporary concept of heritage has been defined, redefined, recalibrated, and officially acknowledged thus gaining a unitary—even global—acceptance and status; most cultures have adhered to international standards that safeguard its presence and ensure its recognition, thereby causing it to become 'a constant of every human group'.[12] Although local and regional differences of the perception of heritage still remain, (first and foremost mirrored in the state of conservation of certain categories of built heritage) the general concept of heritage can be considered universal and homogenous.

The spaces created and used by a community are impregnated with meanings through various practices, and gain plus-value; thus, they become the main support and medium for their cultural identity. The continuous practice assures the 'life' of a particular space; as long as the community perpetuates the specific practices of a space, its cumulated layers of meaning, or its memory can be considered alive. Through this continuous practice all the layers of meaning invested remain interconnected, bridging in their turn all of the generations that have at one time contributed. As contexts change, these spaces—continuously accumulating layers of meaning—become reservoirs of lived memory.

The link between continuous practice and space is essential. When the original practices are no longer performed, these spaces become void shells, no more than imprints. Although the memory of these practices remains imprinted and partially conserved within the material form, these spaces are transformed into places of memory. As the practice vanishes, the physical form either gains value and importance, as a last evidence of a specific context (tradition, community, event, way of life, power-relations, etc.) either becomes neglected or ultimately forgotten. Despite this, the simple preservation of the physical form cannot guarantee its ultimate desideratum—the preservation of a cultural identity. The loss of practice leads to a disconnection from its context and its present community. When the original practices can no longer be 'read', the object loses its meanings (as they become invisible to the present

12 Michel Foucault, "Of Other Spaces," in *Heterotopia and the City: Public Space in the Post-civil Society*, eds. Michiel Dehaene and Lieven De Cauter (Oxon: Routledge, 2008), 18.

reader), and it becomes disconnected from the cultural identity it belonged to, which, in turn, will ultimately affect the physical state of built object. To safeguard these spaces, original practices need to be retrieved, re-enacted, and recreated[13] in the present. A means to resuscitate these places of memory consists of the introduction of heritage practices, with the intent of replacing the missing ones.

When the heritage practices are instated, the original function and practices can be either maintained, partially altered, or replaced; the original material substance is generally conserved and protected, or minimally altered in order to adjust to a new function/practice. The symbolic dimension is usually emphasized: the space is recreated as a heritage object and becomes other to its former self. The new status of the object refocuses its perception mainly through its mnemonic and 'identitary' function. The heritage object becomes other in relation to its context, and its otherness is conventionally accepted and perceived. One frequent outcome of the introduction of heritage status is a conflicting situation: the heritage practices replace the original practices, yet fail to stand-in for them. This abrupt termination of the original and/or ongoing practices has the specific preservationist aim to reduce the functional ware of the physical form, yet the introduction of the status acts similarly to the disappearance of the original practices, as it leaves behind an empty yet protected shell. The heritage object is thus incapacitated, locked in an impasse: it is being denied its original practices while its new protectionist ones are unable to reconcile it with its context. The heritage object becomes other in relation to its previous self and to its context, yet its otherness manifests through isolation, marginalization and ultimately invisibility. If one was to compare these two situations of the introduction of the same (protectionist) ordering, one could conclude that *otherness* has a potential positive implication as well as a negative one.

The contemporary patrimonial gaze—ever-expanding and aware—is oriented towards its narcissistic reflection(s), seeking to achieve through plurality, its continuity and homogeneity. Both Choay and Kovacs state that the contemporary society has come to ultimately define itself through its mirrored image, through its past achievements, (whether 'golden ages' or irretrievable mistakes) hence its obsessive preoccupation with built heritage. Heritage in

---

13 Laura Angélica Moya López and Margarita Olvera Serrano, "La experiencia de la temporalidad en las sociedades contemporáneas, identidades sociales y rituales conmemorativos. Una propuesta de análisis desde la sociología y la teoría de la historia", *Sociológica* 26, no. 73 (2011): 59, apud. Pierre Nora, *Realms of Memory* (New York: Columbia University Press, 1996), vols. 1–3.

general and specifically built heritage are fragments assembled to reflect one's identity, and thus gains their mirror function.

As Choay remarks in an overview of the historical process of heritage-value endowment to the built object (also common to the movable object), the initial 'primeval' phase is not motivated by a connection with a certain historical event or period, or by its age. This first value-status historically given to the built object was constructed as a process of appropriation through which the material manifestation of a different culture, deemed superior, was showcased in order to 'revive an art of living and a refinement that only the Greeks had known'.[14] Also several other adjoining reasons seem to be part of this initial layer of value, such as esthetical pleasure, prestige, snobbery, and profit;[15] although this value status entails to some extent the outlining of basic conservational practices, it is attributed to a very limited class of objects—built or mobile: those belonging to the Greek antiquity.

The ground zero of the monument protected status is the remaining antique built fund. Choay identifies two main reasons behind the deliberate conservation and the conferring of value-status to the built object, both evolving from the 6th up to the 11th and 12th centuries. The first reason had a practical economical nature: the reuse and repurposing of an already existing, although damaged, architectural reserve; the second reason was the 'intellectual attraction … and the seduction of sensibility'[16] that the classical built heritage inspired within the humanistic community. Although the epoch acknowledged the built objects' qualities and conferred it a value-status, its conservation remained one focused on repurposing of the spaces (with new functions) or on the re-use of the materials (as architectural elements or simply as building materials).[17] The process of value bestowing evolves in its second phase, adding a new layer of meaning. The built object is now thought of as a vehicle of memory (a medium of 'the will of asserting one's identity through monuments'[18]). The 'distancing gaze'—that thrived in the following 13th to 15th centuries and on—becomes more defined. The built object becomes the 'testimony of the reality of a completed past',[19] an 'identitary' tool, a 'monument'

14 Françoise Choay, *L'allégorie du patrimoine* (București: Simetria, 1998), 20.

15 Id., 23.

16 Id., 25.

17 Choay entitles these two approaches as 'global reuse [accompanied or not by alterations]' and 'fragmentary reuse … in parts and pieces, used in diverse purposes and places'. Choay, *L'allégorie*, 27.

18 Choay, *L'allégorie*, 29.

19 Id., 31–35. Choay sees in this view towards the built monument two attitudes, the esthetical one and the scholarly one (or the historical perspective: the antique edifices are

that must be preserved and showcased. Despite this acquired value-status, the built object isn't always protected: the safeguarding of edifices is selective and troublesome, raising far more complex technical, economic, social, or urban difficulties than its movable counterpart, the artefact, more easily sheltered in private collections, *studioli*, or galleries. In this phase, the conservational initiatives are circumstantial and there are no established protectionist methodical strategies aimed at safeguarding the built object.

In the following phase, the antiquarian perspective shifts the focus from the written testimonies authors to the object itself: a more tangible and thus reliable source of information. The antique monuments and public utilities will be enlisted and categorized, as they 'appear particularly rich in information to the extent that they represent the spatial frames of institutions'.[20] As this process of acknowledging the cultural and social value of the built antique object unfolds, the temporal and spatial area of interest broadens as well, including the 'formerly ignored, neglected or disregarded cultures'.[21] The built object is thus officially acknowledged in its extensive documentary value, and recognized as an imprint of cultural identity and of specific practices. Choay observes that within this class of built edifices a specific category stands out, due to its symbolic value and its very rich representational or iconographic documentary value—the church; this particular type and its cumulative layers of value will be analyzed further on in the course of this chapter.

The main 'built in' encoding, as well as the layers of meaning acquired in time are often overshadowed by problems of identification and interpretation, and the built object tends to be either erroneously attributed, either misinterpreted (its iconographic elements, its architectural space, form, and function). Advances in the scientific research and comparative analysis redefine the attitude towards its object of study. The 'distancing gaze' already achieved by the 18th century interpolated the lens of a sharply delineated historiography: the material objects are acknowledged as imprints of disappeared cultural practices. The final past/present rupture and the crystallization of the modern historical consciousness claims its main source in the industrial revolution, and the radical shifting of the hierarchy of values that came with it.[22] Furthermore, the

viewed as a confirmation of illustration of the classical texts, yet inferior to them); these two views will eventually merge through 'a process of mutual impregnation', and a primeval and selective notion of 'historical monument' would be born.

20 Id., 46.

21 Id., 47.

22 Choay, *L'allégorie*; Jukka Jokilehto, *A History of Architectural Conservation* (Elsevier Butterworth-Heinemann, 2002), 16–18.

aesthetic dimension comes back into focus as new art-related practices and values are being established and disseminated.[23] The romanticist contemplation of the picturesque gives the built object a new layer of meaning; despite being integrated and staged within a larger context (natural, rural, or urban landscapes) and far from being the main interest of the romanticist gaze, the built inherited object gains emotional and moral value. Both the nostalgia for a glorious past and the experience of the impermanence of the human existence are means employed to arouse an emotional response from the viewer.

Both the historiography and the esthetical approach embody a fundament for protectionist policies. Yet, only the reassertion of the 'identitary' and mnemonic character of the built object (within a nationalistic context) will establish an incipient conservationist structure.[24] Choay sums up these values as follows: *the national value*—built object as an illustration and support of the national feeling; *the cognitive/educational value*, the *economic value*, and the *esthetical/artistic value*.

At the beginning of the 19th century the built object—now read as 'the historic monument'—gains yet another value, one dictated by its uniqueness and irreplaceable nature. The traditional architecture becomes the opposite of the serial, the mechanic, the produced, the calculated, and the anonymous; through this antithesis of the old and the new, the built historical monument, weather public or private, becomes an idealized representation of the permanence of the sacred.[25] Two different approaches caused by the industrialization are typically acknowledged: the French attitude, promoting a double nationalistic-historical value; and the British attitude, through Ruskin's approach, promoting a more integrated, diversified, and subjective value: the domestic architecture, or the architecture of the everyday, is acknowledged its sacred dimension, and its identity-bearing value.[26] Bringing the 'minor' architecture into focus, Ruskin propelled the universal character of the built monument.[27]

These shifting hierarchies of value have given rise to protectionist practices, although sporadic, hesitant, and sectional (the classifying-oriented antiquarian

23 Id., 63.

24 Choay mentions both the British response, that generated the first protectionist structure created in order to safeguard the nation's built heritage, and the French response, emerging more than half a century later.

25 Choay, *L'allégorie*, 100.

26 Id., 101–105. Ruskin offers an interpretation of the historical monument of the everyday, shifting the focus from the general, all-encompassing historical perspective, to the particular histories of the communities, of the local, and of the everyday.

27 Choay, *L'allégorie*, 101–105.

interest, the esthetical and subjectively analytical graphical representations, the written protests against destruction, selective and subjective preservation interventions, etc.). The scale, effectiveness, and diversity of the destructions entailed by the industrial revolution (restorative vandalism, destructive vandalism, and pollution) have completed the value-sifting rift; new practices were needed in order to safeguard the now severely vulnerable historical built fabric. The two main official protectionist practices—a unitary corpus of laws and an interventionist discipline, summing up conservation practices[28]—are to be established, acknowledged, and applied by the end of the 19th century, in most European countries. These protectionist official practices reflect a more structured and controlled approach of the historic monument and of what is to be considered heritage-value; the patrimonial gaze is fully aware of itself and of its subject. The protected heritage status institutes, acknowledges and validates the objects' special nature and its cumulated layers of meaning. As the object gains its protected status, it is officially recognized as *other*.

Through its newly instated ordering—introducing privileges, prohibitions, prescriptions, norms, and regulations on both its own space and its surrounding space—its enclave nature is emphasized. As a direct result from this enclave-like nature a double adjustment of its accessibility—physical and mental—is operated. The heritage space is both isolated (through protectionist restrictions) and permeable (in order to serve its main purpose, as a cognitive/educational tool). The access is within the space controlled: physical and psychological rituals of entry that are designated according to the roles assumed by those entering the space. Even when the delineations and enclosures are invisible, the access is mediated through symbols, announcing its status and its particular functioning.[29]

28 Delafons discusses the linguistic issue brought on by the disciplines and approaches involved in the built heritage domain, responsible for the distortion of the notions defining the protection and intervention on the built object—*restoration, preservation, protection* and *conservation*. '...*restoration* implies significant work on the fabric, which may extend to substantial rebuilding; *preservation* implies retention with minimal alteration; *protection* implies safeguards against demolition or ill-advised improvement but does not exclude possible adaptations or alterations; finally, *conservation* goes much wider than the protection of individual buildings and can extend to whole areas and to features other than buildings. It also implies a different policy approach, reflecting a broader range of public interest'. Delafons, *Politics and Preservation*, 4.

29 The heritage status implies numerous and alternate rituals of entry—depending on the particular space, and the role assumed by the one accessing the protected space—as well as a certain awareness of its status. If such awareness is lacking the *space* may be accessible yet the *place* remains elusive; without acknowledging its cumulated meanings the place remains inaccessible.

The heritage status also initiates a new functioning of the space: the entire normative system is introduced in order to abolish the destructive passing of time. As it is detached from the natural flow of time, for preservation purposes, the space becomes a 'slice of time':[30] at an 'absolute break'[31] with its present, and existing within its own parallel time. It also becomes a showcasing of its 'original time' (of what is introduced as such). The physical form of the object is to be preserved (thus stopping its time flow) and restored (thus reversing its time flow). Within this particular heterotopic feature lies the utopian nature of heritage, as it is driven by the struggle to suspend time and its impossible aim to freeze the physical form of the object.

From the same protectionist status emerges its dual illusory character. Since it functions with the immediate purpose of 'reconstructing [one's] roots, [as] a compensatory fictional space in the past, a *pseudo-topie*, in order to try and artificially recreate the differences that the present no longer admits. The past becomes ... a refuge-value';[32] the protected space thus becomes an illusion of permanence in the face of time. It can be read as both a space of compensation (the golden age, an idealized extinct world of a more real set of values) and a space of illusion—since the idealized retrospective is always selective of its memories. Confronted with its contemporary context, it almost always reveals it as transitory and illusory.

As the heritage gaze has widened, its scope has assimilated both extra-ordinary and infra-ordinary built spaces.[33] Heritage has come to encompass contradictory and mutually exclusive spaces, in both its conceptual aspect and its more practical, physical aspect. Both conceptual and physical aspects of

30 Foucault, "Of Other Spaces," 20.

31 Id., 20.

32 Marc Guillaume, *La Politique du Patrimoine* (Paris: Éditions Galilée, 1980), 15. The original quote: 'A ceux qui n'ont plus ni territoire ni identité sociale propre, la seule possibilité qui reste ouverte est de se reconstruire des "racines", un espace compensatoire fictif dans le passé, une pseudo-topie, pour tenter d'y recréer artificiellement les différences que le présent ne tolère plus. Le passé, comme l'écologie, devient valeur-refuge. Pour briser l'uniformité et le fonctionnalisme du paysage industriel et des logements, pour les rendre habitables, les débris anciens restent le dernier recours.'

33 '...the infra-ordinary built fund, devoid of historical or artistic value, yet performing a specific unique and identitary function that is irreplaceable'; Kovacs specifically refers to the traditional urban fabric 'which continues to disappear rapidly'. Kovacs, *Timpul monumentului istoric*, 62, apud. Georges Perec, "Approaches to What? [1973]", in *The Everyday Life Reader*, ed. Ben Highmore (London and New York: Routledge, 2002), 176–178, accessed December 11, 2017, http://artsites.ucsc.edu/faculty/Gustafson/FILM%20162.W10/readings/perec.approaches.pdf.

heritage succeed in 'juxtaposing several spaces in a single real space'.[34] The built heritage is capable of embodying more than one culture,[35] and protectionist status not only acknowledges but also emphasizes its plural nature.

The heritage space is endowed with these features as protectionist policies and status are introduced. The mirror function, the constancy within every human group, the enclave character, the 'slice of time' character, its compensatory and its illusory characters, its capacity of juxtaposing mutually exclusive spaces, its controlled access—all reflect the heterotopic character of heritage.[36] Thus, the physical as well as the conceptual space of heritage can be read as heterotopic, especially within the contemporary cultural and socio-identitary context.

The Heterotopic Character of the Historical Wooden Churches of Transylvania

Heritage as a heterotopic space can be identified in many instances, one of which being the ecclesiastical architecture of Transylvania, the wooden 'blockbau' shingled church typology. Both its overall silhouette and its building technique are shaped by its environment, its natural or geographical one, as well as its oppressive socio-politic one, which impacted especially on the endurance in time of its building materials.[37] The externally imposed cult and prohibitions influenced the design and the lifespan of these churches, yet they reinforced and involuntarily conserved their original orthodox religion, architecture, and practices. Most of the communities in the region refused catholicization and maintained the religious service in its original byzantine

34 Foucault, "Of Other Spaces", 19.

35 Craig Forrest, *International Law and the Protection of Cultural Heritage* (Oxon: Routledge, 2010).

36 Following the six principles enumerated in Foucault's text, entitled *Des Espaces Autres*, published posthumously by the French journal *Architecture/Mouvement/Continuité* in October, 1984.

37 'The oppressive Austro-Hungarian feudal rule did not allow for the [traditionally Orthodox] Romanians in Transylvania to construct their ecclesiastical buildings in a lasting manner', thus prohibiting the use of stone and masonry, that was to be reserved for the official, accepted religions and nations; this injunction, along with others, served the consolidation of the Austro-Hungarian Empire's authority, through the introduction of a forced Catholicism. Ioan Godea, *Bisericile de lemn din Romania—Nord-VestulTransilvaniei* (București: Meridiane, 1996), 32–33, 128.

ordering (the liturgical acts,[38] ritual, gestures, and motions—the mandatory cross sign, anointing and timed prostrations, the timing structuring of the services in a liturgical day,[39] fixed celebrations, Lents, established emplacements and ritual flows, the specific traditional terminology, etc.). Although the catholic influence is indisputable, the contemporary rural cultural practice maintains mostly unaltered the fixed structure of the traditional religious service.[40] The built object preserves numerous imprints of these practices, in its spatial layout,[41] the prescribed location of mural paintings and placement of icons, the gender spatial separation, even physical imprints of recurring customs.[42]

The perishable nature of these churches was externally imposed through historical specific politics of invisibility, yet these shifted to visibility as the official protectionist ordering was introduced; they acquired a protected status and consequently became markers of cultural identity. The protectionist status also instilled the constant fight against their perishable nature, and, as with heritage objects of a similar nature, against their everyday functional ware through customary practices, that allowed and encouraged alteration of the space.[43] The introduction of official orderings usually challenges and

38 The temporal structure of the service follows the graeco-roman dividing of the day in 4 periods, 'orae' (of 3 hours each), and of the night in 4 periods, 'watches' (3 hours each). Vasile Răducă, *Ghidul creştinului ortodox* (Bucureşti: Humanitas, 1998), 106–109.

39 i.e. 'Liturghia' (ro.) from the Greek '*leitourghia*', is the counterpart of the catholic *mass*. Răducă, *Ghidul*, 109.

40 Other additions are either regional or local, of pre-Christian nature, grafted from the older animistic background; their influence produced an interesting hybrid, mostly expressed through customs and traditions, and in specific/prescribed uses of certain spaces.

41 The tripartite structure reflects the religious practice—the altar apse nave and the narthex are designated spaces for the hidden sacred service, the men's and women's separated spaces, and the neophyte's exterior space.

42 These physical imprints, sometimes registered by restorers, are proofs of the continuous functional ware through ritual, such as candle burn marks in specific areas of the iconostasis or icons, the erasing of the mural painting in specific areas, nailing of crosses or other objects (horseshoes, cloths) to the walls in specific places (lower/upper beams, iconostasis, etc.).

43 The customary solution to the natural growth of the community was the enlargement or replacement of the church with a more spacious one (since the destruction of a church was prohibited on religious grounds, the building was disassembled, and sent, sold or given to another community in need of one); as the heritage status is introduced, any physical alteration is prohibited, allowing as the only option the construction of a new church. This mechanism led to further isolation and the strengthening of the enclave character of the church.

contradicts the traditional ordering. In the case of the wooden churches the traditional ordering is compelled (by legal means) to displace and relocate itself, yet in the case of another traditional space discussed in this book, the official ordering introduced is altered in order to accommodate the traditional ordering, or the everyday traditional practices. This is the case examined by Assumpta Nnaggenda Musana and Eiman Ahmed Elwidaa.[44] The Ugandan low-income housing schemes, based on non-traditional western models, is introduced as an official ordering, yet it eventually becomes altered and tailored to the user's practices and socio-cultural perceptions of space. Contrary to the Romanian regulated heritage space, the residential Ugandan space is unrestrained, at liberty to reshape the official ordering; despite these different approaches, the process is similar: the community adapts the space it inhabits/practices,[45] it imprints itself through practices into the physical from of its spaces. These controlled and highly controlled spaces—through urban planning projects, through architectural design, or through function related standards, such as airports, prisons, etc.—are lived and 'practiced' in an alternative manner, differently from the 'officially' designated mode. These 'other' uses of regulated space shape and adapt it in order to fit the daily 'real' fluxes, hierarchies, and values functioning within that specific community (as opposed to those hierarchies and values that are top-down defined, or externally imposed). In both cases the space becomes 'identary', containing and reflecting the perception of its users towards the space itself as well as towards themselves and their way of life.

In the case of the Transylvanian wooden churches the traditional ordering included interventions such as enlargements or extensions of the space. One quite frequent solution to the natural growth of the community was the replacement of the church with a more spacious or solid one; since the destruction of the old church was prohibited on religious grounds, the building was

---

44 Assumpta Nnaggenda Musana and Eiman Ahmed Elwidaa, "Women as Retrofits in Modernist Low-Income Housing", in this volume.

45 *Pratiquer l'espace* (fr.), or 'to practice a space' as defined by Michel de Certeau (referring to 'la pratique du lieu', or the practice of space) where through the practices, a used *place* becomes *space* (or space becomes a practiced place); also explained by Mathis Stock: 'On peut définir les "pratiques des lieux" rapidement comme étant ce que font les individus avec les lieux, étant entendu que ce sont les manières de pratiquer les lieux qui retiennent notre attention, non la question de la localisation ou la fréquentation.' Mathis Stock, "L'habiter comme pratique des lieux géographiques", *EspacesTemps.net*, December 18, 2004, accessed December 11, 2017, https://www.espacestemps.net/articles/habiter-comme-pratique-des-lieux-geographiques/.

disassembled, and sent, sold, or given, to another community in need of one.[46] As the new ordering—the heritage status—is introduced, the traditional ordering is interrupted, as any physical alteration is prohibited, allowing as the only option the construction of a new church. In a puzzling unfolding, the everyday continuous ordering of the space is denied and replaced by the protectionist ordering. The daily practices and original function are excluded, and the enclave character of the space is further emphasized. The space becomes a contested one, unable to accommodate its traditional practices, as a result of the juxtaposition of its layers of meaning—the initial historical conditioning, as well as their protected one.

Their polarizing function within the settlement is diminished, through introduction of status, as well as through a three-step socio-political reordering (in the 18th and 20th century),[47] reinforcing their isolation, their enclave character, and also their marginality. They also become *hybrid spaces*: the religious-sacred encoding of space is superseded and diminished, as they are transformed into almost-public spaces, with its initial gendered and sacred orderings annulled; the space becomes exposed to the tourist-gaze and touch. This layering or 'heritage-ization' process, common to most of the remaining 17th to 19th century wooden churches of Transylvania, generated different outcomes.

The initial by-design enclave nature and otherness of the wooden churches within their settlements meant that their 'heritage-ization' reinforced their isolation and enclave character, as well as their heterotopic one. Through the following case studies it is illustrated how the 'heritage-ization' process generates heterotopic space, in various instances, within the same architectural category, and yet with different outcomes.

The first two cases—the St Paul the Apostle church of Săliștea Veche[48] (figures 7.1 and 7.2) and St. Archangels Michael and Gabriel church of Sânmihaiu Almașului[49] (figures 7.3 and 7.4) (17th)—are both *marginally located* within the settlement. Both spaces are no longer 'practiced' and they remain in their original placement, relatively isolated from the village. In both cases new

46 Since the church could be sold, the responsibility of its perpetuation was passed on; the buyer could use it as he pleased, even as building material. This is the case of the Săcădat church, sold in 1906, and converted by its buyer into a house. Godea, *Bisericile*, 41.

47 These three phases of reordering represent the resettlement of villages that occurred in Transylvania: the initial 18th century phase, the agrarian reform of 1945, and the national program of 1980–89.

48 Located in Cluj County, Romania; dating 1589–1600; graded CJ-II-m-B-07750 according to the National List of Historical Monuments (LMI) 2010.

49 Located in Sălaj County, Romania; dating 1778–1794; graded SJ-II-m-A-05109, LMI 2010.

FIGURE 7.1 *The isolated St. Paul the Apostle Church of Săliștea Veche, Cluj County, Romania—marginally located within the village*
PHOTO: SMARANDA SPÂNU.

FIGURE 7.2 *The advanced state of decay of the isolated St. Paul the Apostle Church of Săliștea Veche, Cluj County, Romania*
PHOTO: SMARANDA SPÂNU.

and more durable churches were built, well distanced from the old ones. The 'heritage-ization' process has been similar, and both are being considered part of the defining regional wooden church typology. Their main difference, of capital importance, is their state of conservation. The St. Paul' church has been abandoned by the community along with its cemetery, and it currently is in an advanced state of decay. Despite its accessible position and proximity to a larger city, it is partially collapsed and rapidly disintegrating. The community's neglect has made its safeguarding almost impossible.[50] The St. Archangels

50 A similar extreme case of neglect is the wooden church of 'DealuNegru', Cluj county; dated 1765, LMI 2010: CJ-II-m-B-07595. Sharing similarities in location, dating, grading and lack of traditional practices as with the cases previously mentioned, the relatively new

church, sharing a similar position within the settlement, has been restored and maintained, despite being harder to reach from the nearest city. The church has been well maintained by the community and as a consequence, also included in local conservation and media-promotion programs.[51] Both churches gain additional heterotopic qualities through their protected status, yet each have a different evolution.

The centrally located wooden churches reflect a similar situation. Both the 19th century St. Archangels Michael and Gabriel church in Apahida,[52] (figures 7.5 and 7.6) and the St. Archangels Michael and Gabriel church in Baica,[53] (figures 7.7 and 7.8) are enclosed within the settlement's built-fabric. Both have been registered as monuments and consequently restored; in order to conserve

FIGURE 7.3 *The isolated wooden church St. Archangels Michael and Gabriel of Sânmihaiu Almașului, Sălaj County, Romania—marginally located within the village*
PHOTO: SMARANDA SPÂNU.

---

restored church was burned to the ground in 2012. Its flammable material and isolation contributed to the collapse.

51 The local authorities have initiated several interventions: grading of more than 70 churches (of more than 80 in the county), implementation of several restoration programs, designing tourist routes for 7 objectives (in progress), initiating the UNESCO classifying procedures for one objective and locally promoting the county with the logo 'Salaj, the land of churches'. 'Județul bisericilor de lemn n-are loc în patrimoniul unesco', *Adevărul*, April 28, 2011, accessed June 18, 2013, adevarul.ro/locale/zalau/judetul-bisericilor-lemn-n-are-loc-patrimoniul-unesco-1_50ad6ed87c42d5a663950e63/index.html; see also "Raport de Țară Sălaj Lăcașul Bisericilor de Lemn Nepromovate', (June 6, 2013) retrieved from http://www.digi24.ro/stire/RAPORT-DE-TARA-Salaj-lacasul-bisericilor-de-lemn-nepromovate_109019.

52 Located in Cluj County, Romania; dated 1806, graded CJ-II-m-B-07518, LMI 2010.

53 Located in Salaj County, Romania; dated 1645, graded SJ-II-m-A-05011, LMI 2010.

FIGURE 7.4 *The St. Archangels Michael and Gabriel church of Sânmihaiu Almașului, Sălaj County, Romania—in a fair state of conservation (preserving both the vivid interior paintings and original details—the wooden door lock mechanism)*
PHOTO: SMARANDA SPÂNU.

FIGURE 7.5 *The St. Archangels Michael and Gabriel Church of Apahida, Cluj County, Romania, centrally located within the village*
PHOTO: SMARANDA SPÂNU.

their remaining original substance, they are no longer 'practiced'. The local community has erected new larger and more robust churches. In both cases the new constructions have shifted the polarity of the settlement, converting the centrality of the old church into marginality. The old churches conserve the imprints of their former specific practices yet have become hybrid forms of sacred-everyday space. Similar to their marginally located counterparts, the centrally located churches have evolved differently: the St. Archangels of Baica is in a better-preserved state compared to the one in Apahida, despite the fact

FIGURE 7.6 *The St. Archangels Michael and Gabriel Church of Apahida, Cluj County, Romania. In a poor state of conservation (washing of the interior painting and fungal attack) due to water infiltrations*
PHOTO: SMARANDA SPÂNU.

FIGURE 7.7 *The St. Archangels Michael and Gabriel Church of Baica, Sălaj County, Romania—centrally located within the village*
PHOTO: SMARANDA SPÂNU.

that it is also harder to reach and further away from an urban core. The emphasizing of the heterotopic qualities of the space provoked the surge of the community's awareness. On the other hand, the St. Archangels of Apahida has reached a stage of advanced decay and is neglected by its community, undeterred by its acquired status.

The third situation identified is the most interesting one, since it illustrates most objectively the *mirror function* introduced by the heritage status. Here, two contradicting orderings are confronted, and two opposing instances of time are juxtaposed. The perishable and the permanent, the evolving and the stagnant and the acknowledged and the contested—reflect each other as in a

FIGURE 7.8 *The St. Archangels Michael and Gabriel Church of Baica, Sălaj County, Romania, presents a very good state of conservation and maintains many of the original details*
PHOTO: SMARANDA SPÂNU.

FIGURE 7.9 *The old Assumption Church of Bălan-Cricova, in a very good state of conservation; by its side, the new larger and embellished church* (*right: interior painting of the polygonal dome*)
PHOTO: SMARANDA SPÂNU.

mirror. The same uninterrupted cultural practice is disjointedly represented as two mirrored yet different spaces. It is as if one could contemplate simultaneously, side by side, two temporal instances of the church. The contradictory space of the adjoined churches is exposing the *illusory nature* of their material context. The 'no-longer-practiced' space of the old wooden church is mirrored by the stupendous and rich space of the new church, image of a celebration of its practices; as grand and overwhelming one is the other is small and imperfect (figure 7.9).[54]

54 The wooden churches of Cluj County: Crişeni, Frata, Sălişca Vale, Săliştea Nouă, Valea Caseiului; Sălaj County: Bălan Cricova, Chechiş, Romiţa, Chieşd, Sânpetru Almaşului,

Despite being generated by the same cultural practice, the two adjoining churches are radically different through scale, design, silhouette, materials, and relation to their surroundings; they appear as distorted reflections of each other. The temporal and spatial dislocation thus expressed in physical form originates within protectionist policies. The everyday religious practices are prohibited within one space (as required by conservation principles) thus dislocating them and forcibly transferring them to another space, free of political, material, or protectionist constraints, the adjoined church, rampant and exuberant in its use of materials, its plastic expression, size, and site integration (figure 7.10). The official protectionist politics have intervened in the natural flow of the cultural practices in order to conserve their physical imprint and

FIGURE 7.10 *Several examples of the adjoined churches from the Transylvanian territory, all fairly well preserved: St. Nicholas of Romita, the St. Apostles Peter and Paul of Sânpetru Almașului, St. Hierarch Nicholas of Creaca, St. Nicholas of Chechiș all from Sălaj County, Romania*
PHOTO: SMARANDA SPÂNU.

Răstolțu Deșert; Mureș County: Cerghizel, Cuci; Alba County: Arieșeni, Băgău, Gârda de Sus, Pianu de Sus; Bihor County: Bucuroaia, Câmpani de Pomezeu, Căpâlna, Dușești, Fânațe, Tilecuș, Vârciorog, Vălani de Pomezeu, Cociuba Mică; Bistrița Năsăud County: Suplai, Gersa I; Hunedoara County: Birtin, Lunca, Ocișor, Șoimuș, Târnăvița; Arad County: Julița; Maramureș County: Rogoz, Inău, Coruia, Remetea Chioarului, Vălenii Șomcutei, Bicaz, Moisei Monastery; Satu Mare County: Corund; Sibiu County: Apoldu de Jos.

its 'identitary' and mnemonic features. The complete arrest of the first space almost automatically generates the new one. The new church built in the proximity of the old one generates a paradoxical space, as it is simultaneously in compliance with the protectionist laws (requiring termination of ritual/everyday practices in the old wooden church) and breaching of the same protectionist laws (as it often does not respect the distancing regulations).

These spaces, contradictory in every physical aspect are yet two instances of the same cultural practices and mentalities. The wooden churches were traditionally very flexible spaces, constantly re-tailored and adjusted, enlarged, embellished, and enhanced, following the dynamics of both representational and rural spaces; by safeguarding-freezing them, the protectionist policies have encouraged the very mentalities that created these ecclesiastical spaces. Compared to the above-mentioned situations, the centrally and the marginally located churches, these adjoined churches are better preserved. Despite being dwarfed and outweighed by their modern counterparts, these churches became more visible, their communities becoming almost involuntarily more aware of their presence and value. I argue that the general better preservation of the old church is due part to its proximity to the new one (that becomes the new focal point of the community) and part to its legitimizing capacities: by association the new church is more easily appropriated and recognized as a means of perpetuating tradition. Although in many cases the old church is perceived as physically inferior (too small, decrepit, structurally unsafe etc.) and outdated, the community usually shares an emotional bond with it.[55]

Similar Dynamics with Different Outcomes

The phenomenon of the adjoined churches is not exclusive to the Transylvanian space; several other instances have been identified during this research. Most of these present similar evolutions, such as the Sfinții Voievozi / Saints Archangels Michael and Gabriel wooden church of Șerbănești, Suceava county

55 This emotional bond is usually described through memories: when proudly describing the church in 'its better days', 'the way it was back when I was a child', or 'people used to come from three villages away for the liturgy' [on the titular saint's day] or specific memo ries associated with the space of the church—'I remember climbing in the choir balcony as a child [during the religious service] and look closely at the faces of the saints [painted on the walls and ceiling]', as reported by one of the members of the same community that subsidized the building of the new church.

(1778, SV-II-m-B-05652), the Descent of St. George of Cociuba Mică, Bihor County (1715, BH-II-m-B-01136), or the Saints Arch. Michael and Gabriel of Gersa I, Bistrița Năsăud county (1721, BN-II-m-A-01659), among many others.

As previously argued, these exceptional, heterotopic mirrored spaces have been molded by their context, and by the contemporary practices superimposed over their traditional background. These cases indicate that the traditional practice of renewal (figure 7.11) and perpetual improvement is still active, even if displaced. When uninterrupted, a common outcome of this practice is gradual alteration and finally replacement of the older building with its enhanced version. Another instance of these same dynamics is illustrated by the case of the Saint Parascheva wooden church encased in the fabric of Videle town, Teleorman county. Its current (restored) state showcases two instances of the same practices, yet belonging to two different timeframes: the original wooden church and its improved and embellished version, built *around it* in less perishable materials (brick masonry, painted, and plastered). The same cultural practices and mentalities previously discussed have generated this hybrid. However, in this case the practice of renewal and perpetual improvement is visible only due to its deficient conservation.[56] The heritage 'interruption' acts as a mere framing of this process, working against the total disappearance of the more vulnerable and exposed layer.

Similar mechanisms, even though from different cultural paradigms, can be easily located. Zheng Shanshan notes the same practices in the Chinese space, yet here the top-down delivered heritage status and practices are met with resistance: 'villagers would rather decline the honor of the temple ... being listed as an "historic site"' and 'instead of being satisfied with restoring the historic building to its original condition, [they] have felt free to [fund and] build a new temple in sumptuous style'.[57] This attitude, owing to a different past[58]

56 The plastering of the exterior wooden walls would have been one of the first endeavors for the improvement of these sacred spaces, followed by their decoration. A following stage in the pursuit of durability would have been the dismantling of the old church and the building of a new one in stone or brick masonry.

57 She also observes that 'in order to avoid intervention from local government, some villages were even secretly completing all the restoration work in a very short time without notifying anyone'. Zheng Shanshan, "Religious Diversity and Patrimonialization. A Case Study of the Nianli Festival in Leizhou Peninsula, China", *Approaching Religion* 7, no. 1 (2017): 24.

58 Juliane North discusses several aspects of the Chinese Cultural Revolution (1966–76), 'a synonym for the destruction of cultural heritage' and its 'Destroy the Four Olds' campaign (old thought, old culture, old customs, and old habits). Juliane North, "'Make the

and belonging to an altogether different cultural paradigm reflects different 'heritage-community' and 'community-institutional power' interactions, as well as a different understanding of the heritage apparatus. In the Chinese case, the heritage status and its regulations are challenged and, in this case, even repelled, whereas in the Romanian case they are accepted more easily, which would suggest a lesser involvement with the heritage apparatus. The two attitudes share the 'vernacular' feature of renewal and perpetual improvement of religious spaces. However, in the Romanian case of the adjoined hybrid churches this practice is interrupted and partially displaced; both churches co-exist side by side, defining one another and showcasing two subsequent temporal and spatial instances of the same practices. In the Chinese case, this 'heritage interruption' appears as more indefinite, even when a new temple is built; this can largely be accredited to the coordinates of the cultural paradigm, or to the designated material (architectural) expression which has remained to a great extent unchanged despite the technical advances.

Conclusions

Through the heritage policies, the protectionist status has introduced a new heterotopic layer of meaning, enhancing the existing heterotopic nature of Transylvanian wooden churches, with both positive (visibility, awareness, and conservation of its physical form) and negative outcome (invisibility, isolation, and physical deterioration).

Contrary to the 'museification' tendency and the identity crisis subsequent to the 'heritage-ization' process, in the case of the wooden churches of Transylvania, and specifically in the case of the adjoined churches, these heritage politics have maintained and enhanced the spaces' heterotopic qualities. Through the disruption of the natural flow of the built object, displayed in the individual/shared space of the adjoined churches, the practices, mentalities, and traditional orderings have been better conserved, and also fueled; the very human constructing capacities otherwise considered condemned,[59] have flourished and assembled new identitary spaces, whose expression, whether aesthetically pleasing or not, is truer to the traditional practice and mentalities.

Past Serve the Present': Reading Cultural Relics Excavated During the Cultural Revolution of 1972", in *Cultural Heritage as Civilizing Mission. From Decay to Recovery*, ed. M. Falser (Cham: Springer, 2015), 182.

59 Kovacs, *Timpul monumentului istoric*, 116.

FIGURE 7.11 *The mirrored churches of Dragus—archive picture, 1901.* (*Nicolae Iorga, Neamul românesc din Ardeal și Țara Ungurească, vol 1, ed. Minerva, București, 1906, 119.*)

Bibliography

Burra Charter. Australian ICOMOS, 1999.

Choay, Françoise. *L'allégorie du patrimoine*. București: Simetria, 1998.

Delafons, John. *Politics and Preservation—A Policy History of the Built Heritage, 1882–1996*, London: E & FN Spon, an imprint of Chapman & Hall, 2005.

Forrest, Craig. *International Law and the Protection of Cultural Heritage*. Oxon: Routledge, 2010.

Foucault, Michel. "Of Other Spaces". In *Heterotopia and the City: Public Space in the Post-civil Society*, edited by Michiel Lieven and De Cauter, Dehaene, 13–30. Oxon: Routledge, 2008.

Gibson, Lisanne and John Pendlebury. "Introduction: Valuing Historic Environments", in *Valuing Historic Environments*, edited by Lisanne Gibson and John Pendlebury, 1–17. Surrey: Ashgate Publishing, 2012.

Godea, Ioan. *Bisericile de lemn din România—Nord-Vestul Transilvaniei*. București: Meridiane, 1996.

Guillaume, Marc. *La Politique du Patrimoine*. Paris: Éditions Galilée, 1980.

Hetherington, Kevin. *The Badlands of Modernity: Heterotopia and Social Ordering*. London, New York: Routledge, 2002.

Jokilehto, Jukka. *A History of Architectural Conservation*. Oxford: Butterworth-Heinemann, 2002.

"Judeţul bisericilor de lemn n-are loc în patrimoniul UNESCO". *Adevărul*. April 28, 2011. Retrieved from adevarul.ro/locale/zalau/judetul-bisericilor-lemn-n-are-loc-patrimoniul-unesco-1_50ad6ed87c42d5a663950e63/index.html (accessed June 18, 2013).

Kovacs, Kazemer. *Timpul monumentului istoric—The Time of the Historical Monument*. Bucureşti: Paideia, 2003.

López, Moya, Laura Angélica and Margarita Olvera Serrano. "La Experiencia de la Temporalidad en las Sociedades Contemporáneas, Identidades Sociales y Rituales Conmemorativos. Una Propuesta de Análisis desde la Sociología y la Teoría de la Historia." *Sociológica* 26, no. 73 (2011): 47–86.

Nora, Pierre. *Realms of Memory*, vols. 1–3. New York: Columbia University Press, 1996.

North, Juliane. "'Make the Past Serve the Present': Reading Cultural Relics Excavated During the Cultural Revolution of 1972". In *Cultural Heritage as Civilizing Mission. From Decay to Recovery*, edited by Michael Falser, 181–199. Cham: Springer, 2015.

Perec, Georges. "Approaches to What? [1973]". In *The Everyday Life Reader*, edited by Ben Highmore, 176–178. London and New York: Routledge, 2002. Accessed December 11, 2017. http://artsites.ucsc.edu/faculty/Gustafson/FILM%20162.W10/readings/perec.approaches.pdf.

Răducă, Vasile. *Ghidul Creştinului Ortodox*. Bucureşti: Humanitas, 1998.

"Raport de Ţară. Sălaj Lăcaşul Bisericilor de Lemn Nepromovate" (June 6, 2013). Retrieved from http://www.digi24.ro/stire/RAPORT-DE-TARA-Salaj-lacasul-bisericilor-de-lemn-nepromovate_109019 (accessed June 18, 2013).

Romanian National Heritage List 2010 ("Lista Monumentelor Istorice", 2010), available at https://patrimoniu.gov.ro/images/LMI/LMI2010.pdf.

Shanshan, Zheng. "Religious Diversity and Patrimonialization. A Case Study of the Nianli Festival in Leizhou Peninsula, China". *Approaching Religion* 7, no. 1 (2017): 21–31.

Skounti, Ahmed. "De la Patrimonialisation. Comment et Quand les Choses deviennent-elles des Patrimoines?" *Hesperis-Tamuda* vol. XLV (2010): 19–34.

Stock, Mathis. "L'habiter comme pratique des lieux géographiques". *EspacesTemps.net*. December 18, 2004. Accessed December 11, 2017. https://www.espacestemps.net/articles/habiter-comme-pratique-des-lieux-geographiques/.

CHAPTER 8

Understanding Spirituality in Disabled Places: Focusing on Urban Ruins and Decay

Joongsub Kim

Introduction

When we think about spiritual places, we often turn to remote, natural environments. This chapter suggests that even in the midst of urban decay, such as that in Detroit, the US, instances of meaningful spirituality exist in the built environment. Finding these moments in areas of urban decay is not to ignore the misery or malaise afflicting many residents of Detroit or to glorify urban decay in some way. Rather, these moments can be seen as refreshing and can educate us about the meanings of places from different perspectives.

It seems easy to characterize urban decay as something not noteworthy because to many people, urban ruins often symbolize failure, despair, sadness, bad memories, or hopelessness. This chapter suggests different perspectives and argues that, in many respects, studying urban ruins is worthwhile.

The purpose of this study is to explore a spiritual dimension of urban decay by focusing on constructive or positive aspects of urban ruins. This study investigates the attempts of scholars and practitioners, such as architects, designers and artists, community development organizations, community activists, and residents, to shed new, meaningful, or positive light on urban decay. Such attempts may include passive and active interventions. Making a commentary on urban decay via design or writing, for example, is one kind of a passive intervention. Actually building or adding structure to urban ruins or using landscape to 'treat' urban ruins in an aesthetically pleasing way,[1] all involving small-scale projects, belong to an active intervention. In both cases, participants are trying to do something about urban ruins. Their efforts are incremental or piecemeal, but nevertheless are considered constructive or positive because they are searching for positive purposes, lessons, meanings, or unknown potentials in the ruins or because they look forward to bringing about new changes to the urban ruins.

1 Brent D. Ryan and Daniel Campo, "Autopia's End: The Decline and Fall of Detroit's Automotive Manufacturing Landscape," *Journal of Planning History* 12, no. 2 (2012): 95–132.

 | DOI 10.1163/9789004376793_009

This study does not focus on examples where the trace or sign of urban decay or urban ruins is completely disappeared or removed (such as the examples where a place is fully revitalized or rebuilt). People in cities such as Detroit often want all urban ruins to be eliminated, but the reality is that a complete eradication of urban ruins is impractical or impossible. Despite good and honest efforts by many people and organizations in Detroit, urban ruins have been there for many decades. Therefore, it is highly likely that residents of underserved areas in cities such as Detroit will face urban ruins near their homes or neighborhoods on a daily basis. In this regard, one of the difficult challenges, then, is how to deal with those urban ruins until all or the majority of them are cleaned up.

This chapter suggests that exploring spirituality in the places in urban decay is one way to address such a challenge. While this study will look into examples in other cities, the chapter will focus primarily on Detroit, since its decay has received a lot of attention from conventional media and social media, including scholarly publications.

Literature Review

What makes some places in urban decay spiritual? What are the key characteristics of such places? To answer those questions, this study first examined the literature on architecture, culture, spirituality, and phenomenology. Research into past accomplishments in the general areas of spirituality in architecture and religion and spiritual places has been beneficial to this chapter. While there are a number of ways we could study contributions made by the current literature, place typology and its impact on the study of spirituality, scientific rigor in the study of spirituality, and the impact of religion contribute to the author's focus. Particularly, this chapter focuses on the first two contributions, as they are more relevant regarding the research questions.

Place Typologies

The pertinent literature in the general area of spirituality takes into account place typologies as an approach to studying spiritual places. Studies have often focused on spirituality in exotic places with striking natural beauty, in popular places such as well-known tourist attractions, and in well-maintained places with long histories as traditional urban environments.[2] That is not surprising

2 Travis Price, *The Mythic Modern: Architectural Expeditions into the Spirit of Place* (Novato: ORO editions, 2012); Karla C. Britton, ed., *Constructing the Ineffable: Contemporary Sacred*

in many aspects, but the recent rise in out-of-state tourists' interest in urban ruins in cities such as Detroit is somewhat surprising and interesting to note. Many organizations in Detroit provide tours of Detroit ruins.[3] They include guided tours of industrial ruins, neighborhood ruins, and other types of urban ruins in the city. Tourists participating in these tours come from as far away as Europe. Some visitors may be motivated to come due to their 'morbid curiosity' about Detroit's malaise, but others come because of their intellectual curiosity or commercial interest, as some of the ruins have significant historic value or are on the market for sale. Others visit these urban ruins because they are attracted to their potential, as evidenced by a recent surge in the number of proposed commercial developments in the places in urban decay. For example, an upscale Whole Foods Market retail store was opened in midtown Detroit in 2014.[4] In particular, midtown Detroit has received extensive media coverage in recent years.

Despite growing intellectual, spiritual, or commercial interest in urban decay and its impact, which may be potentially positive, the literature in spirituality rarely focuses on urban ruins, especially in postindustrial cities like Detroit that have been experiencing shrinkage for decades.

In the current literature on spirituality, studies of place typologies are based primarily on tourist attractions. There is a need to expand the typologies to include urban ruins. A place typology approach informs us of the significance of understanding the unique physical characteristics of a given place (e.g., place attributes) and various ways in which it impacts people who live, work, or visit there (e.g., their activities and conceptions about place).[5] Here, Canter's place model, though not new, is useful to a place typology approach.[6] This place model advances the study of spirituality by looking at it more in terms of the perspective of human behavior and also the relationship among a spiritual place, human behavior, and perceptions associated with it. These behavioral and perceptual dimensions will be further discussed in the Results section.

Architecture (New Haven: Yale School of Architecture Press, 2010); Rubio Eduardo, *Holy Places, Sacred Sites: A Journey to the World's Most Spiritual Locations* (London: Thames & Hudson, 2008); Thomas Barrie, *Spiritual Path, Sacred Place: Myth, Ritual, and Meaning in Architecture* (Boston: Shambhala, 1996).

3 See for example, http://dhivedetroit.org/tours/.

4 Nathan Bomey, "Whole Foods Sets June Opening for Detroit Store," *Detroit Free Press*, March 1, 2013.

5 Mahyar Arefi and Menelaos Triantafillou, "Reflections on the Pedagogy of Place in Planning and Urban Design," *Journal of Planning Education and Research* 25, no. 1 (2005): 75–88.

6 David Canter, *The Psychology of Place* (London: Architectural Press, 1977).

Scientific Rigor

Another achievement by other studies in the area of spirituality is the advancement of scientific investigations of spirituality in architecture; for example, studies by Sterrett and Thomas[7] and other scholars (e.g., see studies by Britton and Todres[8]), as well as various works, especially in the field of phenomenology (e.g., see studies by McGrath and Wierciński; Tilley; Seamon; and Seamon and Mugerauer[9]), help conduct a more robust study on spirituality. Some of the recent studies have advanced the general areas of spirituality in architecture and religion and spiritual places particularly through more grounded study (e.g., see studies by Barrie and Mugerauer[10]), systematic inquiry (e.g., see Todres' study[11]), and empirical investigation (e.g., see Britton's work[12]).

While the newest edition of *Architectural Research Methods* by Groat and Wang[13] does not focus on research in spirituality, the book has expanded discussion on case studies, interpretive-historical research, qualitative research methods, and combined strategies. Some of these methods are informed by some of the recent advancements made in the general areas of spirituality and religion in architecture and phenomenology. In particular, a more increased use of combined strategies and cross-disciplinary approaches in some of the spirituality-related studies (e.g., studies that explore the deep connection between religion and spiritual places, such as studies by Bergmann and

7 Joseph Sterrett and Peter Thomas, ed., *Sacred Text, Sacred Space: Architectural, Spiritual, and Literary Convergences in England and Wales* (Boston: Brill, 2011).

8 Karla C. Britton, ed., *Constructing the Ineffable: Contemporary Sacred Architecture* (New Haven: Yale School of Architecture Press, 2010); Les Todres, *Embodied Enquiry: Phenomenological Touchstones for Research, Psychotherapy, and Spirituality* (New York: Palgrave Macmillan, 2007).

9 Sean J. McGrath and Andrzej Wierciński, ed., *A Companion to Heidegger's Phenomenology of Religious Life* (Amsterdam and New York: Rodopi, 2010); Christopher Tilley, *Body and Image: Explorations in Landscape Phenomenology 2* (Walnut Creek: Left Coast Press, 2008); Christopher Tilley, *A Phenomenology of Landscape: Places, Paths, and Monuments* (Providence: Berg, 1994); David Seamon, ed., *Dwelling, Seeing, and Designing: Toward a Phenomenological Ecology* (Albany: State University of New York Press, 1993); David Seamon and Robert Mugerauer, ed., *Dwelling, Place, and Environment: Towards a Phenomenology of Person and World* (Boston: Kluwer Academic Publishers, 1985).

10 Thomas Barrie, *The Sacred In-Between: The Mediating Roles of Architecture* (London and New York: Routledge, 2010); Robert Mugerauer, *Interpreting Environments: Tradition, Deconstruction, Hermeneutics* (Austin: University of Texas Press, 1995).

11 Todres, *Embodied Enquiry.*

12 Britton, *Constructing the Ineffable.*

13 Linda Groat and David Wang, *Architectural Research Methods* (New York: Wiley, 2013).

Barrie[14]) is noteworthy and encouraging, as it can attract more collaboration and more cutting-edge strategies to studying spirituality in architecture. The combined and cross-disciplinary methods inspired the author's present study.

Despite these valuable lessons and accomplishments (e.g., location-focused research, place typologies, scientific rigor, the impact of religion, etc.), spirituality literature has neglected spirituality in places in urban decay. While studies on the general areas of spirituality in architecture and religion and spiritual places are beneficial to the author's research, emerging studies in the field of anthropology, landscape urbanism, urban ethnography,[15] environmental psychology,[16] social entrepreneurship,[17] public interest design,[18] design activism,[19] public health, and eco-tourism are also noteworthy and useful for this chapter. By way of cross-disciplinary and combined strategies, these fields can help us examine urban decay with a fresh outlook and draw new and interesting lessons.

Method

To address the research questions posed earlier, this study conducted in-depth interviews with people with long-term familiarity with places of spirituality in Detroit's areas of urban decay, participatory action research with focus groups, and an in-depth physical investigation of place. Many of these participants of this study are longtime (e.g., 20 to 50 years) residents of Detroit. This approach also uses the place model proposed by Canter, which holds that place

14 Sigurd Bergmann, ed., *Theology in Built Environments: Exploring Religion, Architecture, and Design* (New Brunswick: Transaction Publishers, 2009); Barrie, *Spiritual Path, Sacred Place.*

15 See for example Ocejo's work in urban ethnography: Richard Ocejo, ed., *Ethnography and the City: Readings on Doing Urban Fieldwork* (New York: Routledge, 2012).

16 See for example Kopec's work in environmental psychology: Dak Kopec, *Environmental Psychology for Design* (New York: Fairchild Publications, 2012).

17 See for example David Bornstein, *How to Change the World: Social Entrepreneurs and the Power of New Ideas* (Oxford: Oxford University Press, 2007); Alfredo Brillembourg et al., *Beyond Shelter: Architecture and Human Dignity* (New York: Metropolis Books, 2011).

18 See studies by Bell and Wakeford and Bell in public interest design: Bryan Bell and Katie Wakeford, ed., *Expanding Architecture: Design as Activism* (New York: Metropolis Books, 2008); Bryan Bell, ed., *Good Deeds, Good Design: Community Service through Architecture* (New York: Princeton Architectural Press, 2004).

19 See Fuad-Luke's work in design activism: Alastair Fuad-Luke, *Design Activism: Beautiful Strangeness for a Sustainable World* (New York: Routledge, 2009).

emerges from the complex combination of physical attributes, human behavior, and people's conceptions.[20] His model is useful for understanding the way the study participants view place or act on it (e.g., fixing it); for example, participants in the author's preliminary interviews often talked about how places affected them or other people, especially their worldviews, ideas, work, and activities.

After the literature review, the author conducted face-to-face interviews with residents who are familiar with urban ruins in Detroit that have received much attention from social and traditional media. In total, 22 people participated in the interviews. Interviews included both open-ended and close-ended questions, such as 'What kind of emotional reaction do you have when you see the urban ruins?' Based on the results of the interviews, 14 people participated in three focus groups (four or five participants in each focus group). The focus groups were also used to examine or clarify the interview responses. Students at the Detroit Studio—a community outreach program of Lawrence Technological University's College of Architecture and Design, where the author teaches—conducted several site observations, focusing on the surrounding contexts, condition of the sites, and pedestrian and vehicular circulation.

The results of the interviews and focus groups were transcribed and grouped according to several distinct categories. Afterwards, recurring themes across the outcomes of the interviews, focus groups, and site observations were identified.

Results

This section presents the core results of the study based primarily on the outcomes of the interviews and focus groups. Also, findings from the pertinent literature are used to support the statements made by the study participants.

After reviewing the participants' various reactions to the urban ruins, their responses were grouped according to ten categories. After going through several rounds of grouping, the author found that four partially overlapping themes run through the stories told by the study participants.

Spirituality of Urban Ruins

Regarding how the study participants feel about the urban ruins that are presented to them, they are split between positive and negative emotions.

20 Canter, *The Psychology of Place.*

The positive responses typically include a sense of hope, while the negative responses usually include feelings of sadness. A number of people talked about the spiritual dimension of urban ruins, but senses of despair, disappointment, and even anger were also mentioned.

In response to the question of why the urban ruins may evoke a spiritual feeling, the following comment typifies the feeling that some ruins are spiritual: 'Some of these places have enduring interesting qualities.... They inspire me, my neighbors, or other people to do something nice for their community, whatever it may be.'[21] Another study participant stated that:

> They [urban ruins] encourage us to fix them, improve our homes and playgrounds, or clean up the streets in our community.... It is almost as if these ruins yell right at you like 'do something!' ... It is like they wake you up in some way.[22]

Another study participant commented:

> Some of the ruins have become a kind of shrine to me and Detroiters ... like sacred places.... I used to go there and pray that things will get better.... They are also places that bring me a lot of memories.... You know these places may deteriorate rapidly and eventually disappear any time ... but they will remain in my mind forever as long as I live.... You feel attached to some of these places.[23]

A common thread that runs through the participants' comments on many of the urban ruins is the dimension of spirituality. Moreover, several themes that cut across their statements include senses of hope, energy, and empowerment; senses of community, collective responsibility, and permanence; sacredness and religious experience; senses of place and attachment; ethics; and soulfulness, among others.

In response to the question of why certain urban ruins are spiritual or how they can be spiritual, four reasons provided by the participants stood out more strongly than others did: catalytic, therapeutic, reflective, and engaging. While there are overlaps among them, they dominate the study participants' responses.

---

21 Study participant 1, interview with the author, May 20, 2012.

22 Study participant 2, interview with the author, May 20, 2012.

23 Study participant 3, interview with the author, May 20, 2012.

Catalytic

The participants feel that some of the urban ruins have an element of spirituality because they feel that those places motivate people to take action, such as improving urban decay, making a political or social statement about it, or bringing awareness of potentials of the ruins to local residents, politicians, community groups, or civic leaders. One participant stated that:

> North End community residents got together with college students and other supporters and help artists to create [a mural] on a large empty building on the main street of the community.... The mural spans the [building's nine stories] with blast of colors that stream down from the sky like falling tears [symbolizing] a growth of new community, community empowerment, a spirit of hope, sense of community, sense of unity, sense of change, and sense of future.[24]

Another participant commented that:

> Strong graphic images of the ruins warn people about safety, educate them about public health and welfare, and constantly provoke us about doing something, not just talking and complaining about the decay.[25]

According to a number of the participants' comments, many of the urban ruins, especially well-known ones, are spiritual because they can potentially transform people's behavior to bring about positive changes, such as improving the undesirable situations and collaborating with others on art, installation, and design projects. Currently, there are several new projects planned or implemented by grassroots groups and by individual residents or artists in the North End community and other underserved areas in Detroit. Awan et al., in their book *Spatial Agency: Other Ways of Doing Architecture*,[26] and other scholars[27] illustrate a number of examples where everyday people take a significant risk to transform their hopeless situations into more desirable environments, which requires a tremendous leap of faith for a spiritual journey to an uncharted territory.

---

24 Study participant 4, interview with the author, June 9, 2012. The interviewee refers to the mural in figure 8.1.

25 Study participant 5, interview with the author, June 9, 2012.

26 Nishat Awan, Tatjana Schneider and Jeremy Till, *Spatial Agency: Other Ways of Doing Architecture* (New York: Routledge, 2011).

27 See for example Andres Lepik, *Small Scale, Big Change* (New York: The Museum of Modern Art, 2010).

FIGURE 8.1 *'Illuminated Mural 101', Art Work by Katherine A. Craig in Detroit, the US* © JOONGSUB KIM.

Therapeutic

A number of study participants stated that many of the urban ruins they are very familiar with are spiritual because they are or can be a place for consoling themselves, their family members, their friends, or their neighbors. Some of these places have become shrines of sorts, places where residents pay a visit and reminisce, meditate, or pray. Some of the places have disappeared as a result of severe deterioration or vandalism (e.g., scrapping and arson) or lack of visitation by people because of safety concerns. One of the participants commented that:

> When the weather is nice, especially after my church services, I go to a block which [used to be] a community center, not far from where I grew up.... This is a place where my grandparents used to take me to various community stores.... They have been empty for many years ... but not too long ago, residents created a pretty flower garden in an empty lot near the stores.... You can sit near the garden and watch beautiful flowers.... It is a place for healing for me and many others.[28]

28 Study participant 6, interview with the author, July 15, 2012.

Another participant who occasionally visits Michigan Central Station stated, 'A new massive garden created in front of the empty building comforts visitors.... It is like healing the old large scar of the city.'[29] Michigan Central Station (also known as Michigan Central Depot or MCS) opened in 1913 but closed in 1988. At the time of its construction, it was the tallest rail station in the world (see figure 8.2).[30]

Other participants mentioned various other urban ruins that were considered places for healing and consoling. It is interesting to note that many of these places have a green component, such as a garden or a park. Landscape Urbanism,[31] one of the popular forms of contemporary urbanisms, and

FIGURE 8.2 *Michigan Central Station, the US*
© CHRISTOPHER BARTHOLOMEW.

29 Study participant 7, interview with the author, July 15, 2012.

30 Kelli B. Kavanaugh, *Detroit's Michigan Central Station* (*Images of America*) (Charleston: Arcadia Publishing, 2001).

31 Charles Waldheim, *The Landscape Urbanism Reader* (New York: Princeton Architectural Press, 2006); Ryan and Campo, "Autopia's End," 95–132.

research on natural environment in environmental psychology[32] and landscape architecture[33] have received much attention from scholars because of spiritual, therapeutic, or healing effects of green landscape on urban residents.

Reflective

A number of study participants shared their views that the urban ruins can be spiritual because they help reflect on their lives—past, present, and future—and their communities.

A key difference between the therapeutic dimension and the reflective dimension is that the former focuses more on the healing and consoling aspects of spirituality. Although there may be overlap between the two, the therapeutic process is more active than the reflective process. One participant stated:

> I participated in free tours of Detroit's famous ruins organized by nonprofit groups in Detroit.... It is a nice way of re-discovering your own community and your city.... It is a journey that takes me to my root.... I also see a glimpse of [a] better future of Detroit and my community.... Despite all the bad things and tragedies [that have] happened, I feel the positive spirit and hope of [the] city is there.... When I tour it ... the tour tells a story, which makes me think about many things.[34]

Another participant commented:

> My kids and I attended a tour of Detroit hosted by a community organization, which included ... urban ruins, urban gardens in vacant properties, and other revitalized places.... Through this kind of tour, we learn about where we are from and where we are heading ... and changes, both good and bad.... It is educational and also inspiring.[35]

A number of participants talked about the benefits of having a 'story-telling' tour where they learn about the city's past, however unfortunate or tragic it

---

32 Kopec, *Environmental Psychology for Design.*

33 See for example Caroline Constant, *The Modern Architectural Landscape* (Minneapolis: University of Minnesota Press, 2012); Mary C. Fenton, *Milton's Places of Hope: Spiritual and Political Connections of Hope with Land* (Burlington: Ashgate, 2006); Rebecca Krinke, ed., *Contemporary Landscapes of Contemplation* (New York: Routledge, 2005).

34 Study participant 8, interview with the author, August 7, 2012.

35 Study participant 9, interview with the author, August 7, 2012.

may be to some people, and also learn the importance of moving forward. A 're-discovery' type tour of Detroit's ruins was mentioned by many participants as a useful way to reflect on the spiritual aspects of their cultural heritage and how it can help them bring about new desirable changes to the city (figure 8.3).

Scholars, who use story-telling,[36] participatory action research,[37] and urban ethnography[38] in the fields of urban planning, sociology, and anthropology, have studied spiritual journeys and stories, told by local people, who used them to re-connect to their roots in terms of both time and place,[39] pursue reflection in action, and find new meanings in life or in the current environment.

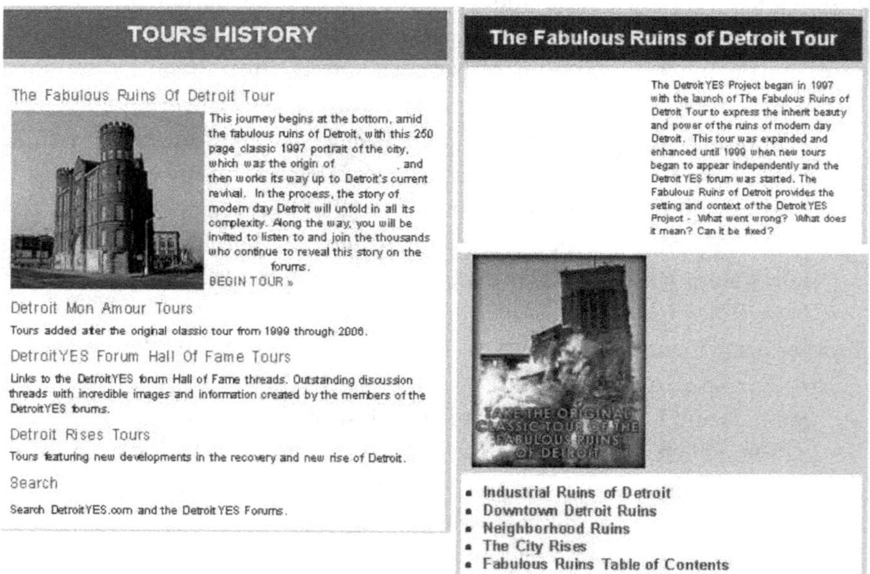

FIGURE 8.3 *'Fabulous Ruins of Detroit Tour', screen shots from web site DetroitYES!*
NOTE: THE SCREEN SHOTS ARE FROM WEB SITES: HTTPS://WWW.DETROITYES.COM/FABULOUS-RUINS-OF-DETROIT/INDEX.PHP AND HTTPS://WWW.DETROITYES.COM/TOUR-DETROIT.PHP

36 John Forester, *Dealing with Differences: Dramas of Mediating Public Disputes* (Oxford: Oxford University Press, 2009).

37 See for example studies by Reason and Bradbury-Huan and Fry in action research: Peter Reason and Hilary Bradbury-Huan, *The Handbook of Action Research* (London: Sage Publications, 2006); Tony Fry, *Design as Politics* (Oxford: Berg Publishers, 2011).

38 Ocejo, *Ethnography and the City.*

39 Dolores Hayden, "The Power of Place: Claiming Urban Landscapes as People's History," *Journal of Urban History* 20, no. 4 (1994): 466–485.

Engaging

Many participants talked about their perspectives on an 'engaging' aspect of spiritual urban ruins. While the reflective dimension connects individuals to their inner selves, the engaging dimension connects them to other people. The engaging dimension focuses on the social aspect of urban decay (figure 8.4). One participant made a comment that especially reflected a sentiment shared by many others:

> If there is one thing that many urban ruins we have in our city, big or small, famous or not-well known, have done to the city, it is the fact that they have brought people together ... not just residents but also outsiders, college students, volunteers, community organizations, philanthropic groups, and even tourists from other US states and countries.... They come to our community, places of urban decay.... One thing they all share is that they want to do something about urban decay.[40]

FIGURE 8.4 *A view of Detroit downtown and the Renaissance Center Building (a symbol of prosperity) and people gathered near the site of a famous urban ruin Globe Trading Company Building*
© CHRISTOPHER BARTHOLOMEW.

40 Study participant 10, interview with the author, September 17, 2012.

Another participant stated:

> Not all engagements result in real changes or successes but there is something special about these urban ruins.... They connect people from all walks of [life] ... connected thorough warm heart and spirit.[41]

There is extensive literature in the fields of environmental psychology,[42] natural resources and environment, and community psychology on the benefits of social interaction and civic engagement and how such activities can lead to helping behavior and social support. In particular, Kaplan and Kaplan[43] and other landscape scholars have written about the benefits of nature, greens, open lands, parks, and gardens (e.g., the positive impact of the spirituality of such places on people's social interactions and reasonable personal behavior).

Discussion and Conclusion

In summary, based on the results of this research, this chapter suggests that the spirituality of places in urban decay can be defined in four ways: Some places in urban decay can be spiritual because they inspire people to do good things for others and the community (catalytic); places in urban decay can be spiritual because they are consoling (therapeutic); places in urban decay can be spiritual because they help connect individuals to their inner selves (reflective); and places in urban decay can be spiritual because they connect people in many different ways (engaging).

One of the key implications of this study is that the places in urban decay can be spiritual and can also have positive effects on people. While other places that are fully revitalized and significantly improved can be equally—if not more—spiritual and positive, the author suggests that studying the ruins that are spiritual and constructive is equally beneficial to cities such as Detroit that suffer from extensive shrinkage, vacant land, deterioration, and other chronic urban ills, because it will take an enormous amount of time, effort, and resources (i.e., long-term, major projects) to undo the damage done to the city over the last several decades. In other words, it can be argued that

41 Study participant 11, interview with the author, September 17, 2012.

42 Kopec, *Environmental Psychology for Design.*

43 Rachel Kaplan and Stephen Kaplan, "Bringing Out the Best in People: a Psychological Perspective," *Conservation Biology* 22, no. 4 (2008): 826–829.

what the city needs urgently is short-term, incremental, and grassroots-level interventions. Many of the urban ruins that are cited by the participants in this study are the ones involving small-scale, piecemeal changes that are spiritual. Often, they are not government-directed but citizen-initiated—not always sophisticated to the eyes of seasoned experts, but nevertheless empowering, engaging, and inspiring.

While the literature in the general areas of spirituality in architecture and religion supports the therapeutic and reflective aspects of spiritual places more than the engaging aspects of spiritual places, the literature neglects the catalytic aspect. The results of this study suggest that the idea of spirituality in architecture needs to be expanded in postindustrial society, in which the old or traditional built environment will continue to decay and in which some residents find spirituality in places in urban decay during their daily lives.

While there are negative aspects (e.g., pessimism) of spirituality in urban decay, the present research also finds many constructive effects of spirituality in urban decay. This chapter suggests that both aspects of spirituality and the role that spiritual places in urban decay play in place-making, especially in shrinking cities such as Detroit, deserve further scholarly attention.

Some of the study participants in this research also shared conflicting emotions about the urban ruins, because sometimes they bring up both good and bad memories simultaneously and also because those memories can be personal (e.g., involving their families or friends in past unfortunate incidents). Furthermore, studying the spiritual aspects of urban decay can be tricky or risky, especially in cities such as Detroit, where residents are still struggling to survive. Despite signs of recovery, albeit weak, the jobless rate in many of the challenging communities remains far above the national average. In the meantime, the crime rate remains among the highest in the nation, and the physical environment, including urban ruins, continues to deteriorate. These are some of the challenges faced when one studies the spirituality of places in urban decay. Perhaps cross-disciplinary research via collaboration among architecture and other disciplines might be one way to address those challenges.[44] It is especially noteworthy that eco-tourism, social entrepreneurism,[45] guerrilla urbanism,[46] grassroots design movement, landscape urbanism, public health,

44 See for example Joseph Juhasz, "The Place of the Social Sciences in Architectural Education," *Journal of Architectural Education* 34, no. 3 (1981): 2–7.

45 See Emily Pilloton, *Design Revolution: 100 Products That Empower People* (New York: Metropolis Books, 2009).

46 See Andrew Smith, *Events and Urban Regeneration* (New York: Routledge, 2012); Tony Fry, *Design as Politics* (Oxford: Berg Publishers, 2011).

public interest design, ethics,[47] and place-making[48] are receiving growing interest from scholars, policymakers, and practitioners in design fields. The lessons from and accomplishments of those fields can help shed new light on the spirituality of places in urban decay.

Additional useful discussions are possible when the author's proposed concept of spirituality in urban decay is compared to ideas discussed in other chapters of this volume. For example, June Jordaan suggests that the spatial characteristics of place can be grouped into three categories: tangible versus intangible space, infinite versus finite space, and meaningful and embodied space.[49] Among these three categories, 'meaningful space' is most useful to understanding significance of spirituality in urban decay in Detroit. Judging from the four domains of spirituality in urban decay corroborated by the participants' comments, as discussed earlier, one can suggest that spiritual urban ruins are meaningful spaces to residents in Detroit, especially to those who have been very familiar with the ruins for many years. This is because they have experienced or have searched at least once in their lifetimes for various meanings in urban decay defined by their own ways, regardless of whether urban ruins are perceived to be tangible or intangible spaces; infinite or finite spaces or not.

In recent years, there have been extensive efforts in Detroit by community groups and non-profit organizations to reactivate some of the vacant buildings in areas experiencing urban decay. These are mostly neighborhood shops that have been closed down for many years or even decades. Usually these efforts focus on temporary reactivation of spaces mainly to bring people's attention to urban ruins, to send a message to the public that they should do something about blight in the middle of the urban neighborhood, and to demonstrate to the residents what (interventions or changes) is possible in spite of chronic urban decay. This has been labeled as pop-up retail projects located mostly in blighted major commercial corridors in the inner cities. Pop-up architecture is a name coined by others and it has appeared in blighted sections of cities across the United States and other countries around the world. Tuuli

47 See Fox's study in ethics: Warwick Fox, *Ethics and the Built Environment* (New York: Routledge, 2000).

48 See Mahyar Arefi and Menelaos Triantafillou, "Reflections on the Pedagogy of Place in Planning and Urban Design," *Journal of Planning Education and Research* 25, no. 1 (2005): 75–88.

49 June Jordaan, "Architectural Agency and 'Place-Making' in a Transformative Post-Apartheid South African Landscape," in this volume.

Lähdesmäki's chapter in this volume discusses the similar examples in Europe. European examples are similar to those in Detroit in two ways: (i) in terms of the goal of increasing people's possibilities of spending time in the city center and, as Lähdesmäki puts it, 'bring[ing] people into less-used or unused urban spaces, and enabl[ing] different people and groups to encounter each other within them'[50] to bring the inner city to life; and (ii) in the way that both examples, as Lähdesmäki describes it, revitalize the less used, unused, or decayed city space by using them as venues for diverse cultural activities.[51]

Furthering the discussion above, more commonality between the two cases (pop-up retails sites in Detroit and pop-up architecture in Europe) can be drawn. Although Lähdesmäki's chapter does not deal with Detroit's cases, what the two cases have in common is 'the use of temporary architectural intervention into city spaces, that is setting up non-permanent structures, constructions, or buildings'.[52] What is different between the two examples, however, is that the European initiatives for pop-up architecture focus more on cultural uses, while Detroit's pop-up architecture focuses on retail. According to Lähdesmäki, the EU's urban initiatives such as

> the annual designation of the European Capital of Culture (ECOC) ... encourages European cities to promote local and economic development through diverse activities and urban projects.... Moreover, the limited length of the ECOC designation (one year) has led the cities to implement and host various small-scale temporary architectural projects in which city spaces are altered through various tactics to achieve diverse cultural, social, political, and economic goals.[53]

In contrast, pop-up retail sites in Detroit have no designation of, or no limit on the time period, location, and use for the temporary architecture or urban development.

The pop-up or temporary architecture in Detroit may be more relevant to the catalytic dimension, one of the four dimensions of spirituality in urban decay discussed earlier. The catalytic dimension includes those places that motivate people to take action, such as improving urban decay, making a political

50 Lähdesmäki, "Temporary Architecture as a Means in Urban Regeneration," in this volume.

51 Ibid.

52 Ibid.

53 Ibid.

or social statement about it, or bringing awareness of the potential of the ruins to local residents, politicians, community groups, or civic leaders. The pop-up architecture helps make those 'catalytic' actions or events occur.

Some of the participants in the author's study of spirituality of urban ruins in Detroit are ambivalent about whether those ruins should be removed (for safety concerns, for example) or protected (for political or cultural purposes). A similar conflict is debated in Smaranda Spânu's chapter.[54] In particular, Spânu debates the controversies surrounding the protection of the historical wooden churches of Transylvania, Romania. Spânu argues that the 'protection targets the conservation of imprints of most often disappeared or threatened practices onto a material form—this being in itself a practice of experiencing the space.'[55] In a similar vein, the preservation of some of the urban ruins in Detroit is a ritual of experiencing the space, by way of a nostalgic gesture to or yearning for a place of bygone era.

Spânu further contends that 'they [protected churches] have become hybrid spaces—both sacred spaces of ritual and free-access desacralised public-spaces.'[56] Similarly, some of the urban ruins in Detroit are considered both sacred spaces of ritual and free-access desacralized public-spaces, yet the protected churches in Spânu's chapter are more likely to survive, in contrast to the ruins in Detroit which will eventually disappear.

As illustrated above, the study of the spirituality of disabled places in urban decay has the potential to contribute to expanding the boundaries of conventional debate in space and place, despite the fact that this study is focused only on Detroit's urban ruins.

This chapter is based primarily on the results of interviews and focus groups of small samples in addition to a literature review. In the follow-up research, the author seeks to include the outcomes of additional interviews, focus groups, and site investigations. This second phase of this research will be conducted as a comparative case study, where the examples of Detroit are compared to those in other cities, to examine similarities and differences among them and to draw useful lessons. The future study will also investigate specific examples of more effective or innovative cross-disciplinary research on the spirituality of places in urban decay. In spite of the limitations of this current study, it contributes in a unique way to research of time, space, and place, in particular in the field of spirituality in architecture.

---

54 Smaranda Spânu, "The Heterotopic Nature of the Built Heritage. The Sacred Wooden Architecture of Transylvania and Its Practices," in this volume.

55 Ibid.

56 Ibid.

Bibliography

Arefi, Mahyar and Menelaos Triantafillou. "Reflections on the Pedagogy of Place in Planning and Urban Design." *Journal of Planning Education and Research* 25, no. 1 (2005): 75–88.

Awan, Nishat, Tatjana Schneider, and Jeremy Till. *Spatial Agency: Other Ways of Doing Architecture*. New York: Routledge, 2011.

Barrie, Thomas. *Spiritual Path, Sacred Place: Myth, Ritual, and Meaning in Architecture*. Boston: Shambhala, 1996.

Barrie, Thomas. *The Sacred In-Between: The Mediating Roles of Architecture*. London and New York: Routledge, 2010.

Bell, Bryan, ed. *Good Deeds, Good Design: Community Service through Architecture*. New York: Princeton Architectural Press, 2004.

Bell, Bryan and Katie Wakeford, ed. *Expanding Architecture: Design as Activism*. New York: Metropolis Books, 2008.

Bergmann, Sigurd, ed. *Theology in Built Environments: Exploring Religion, Architecture, and Design*. New Brunswick: Transaction Publishers, 2009.

Bomey, Nathan. "Whole Foods Sets June Opening for Detroit Store." *Detroit Free Press*. March 1, 2013.

Bornstein, David. *How to Change the World: Social Entrepreneurs and the Power of New Ideas*. Oxford: Oxford University Press, 2007.

Brillembourg, Alfredo, Hubert Klumpner, Patrick Coulombel, and Marie Aquilino. *Beyond Shelter: Architecture and Human Dignity*. New York: Metropolis Books, 2011.

Britton, Karla C. ed. *Constructing the Ineffable: Contemporary Sacred Architecture*. New Haven: Yale School of Architecture Press, 2010.

Canter, David. *The Psychology of Place*. London: Architectural Press, 1977.

Constant, Caroline. *The Modern Architectural Landscape*. Minneapolis: University of Minnesota Press, 2012.

Eduardo, Rubio. *Holy Places, Sacred Sites: A Journey to the World's Most Spiritual Locations*. London: Thames & Hudson, 2008.

Fenton, Mary C. *Milton's Places of Hope: Spiritual and Political Connections of Hope with Land*. Burlington: Ashgate, 2006.

Forester, John. *Dealing with Differences: Dramas of Mediating Public Disputes*. Oxford: Oxford University Press, 2009.

Fox, Warwick. *Ethics and the Built Environment*. New York: Routledge, 2000.

Fry, Tony. *Design as Politics*. Oxford: Berg Publishers, 2011.

Fuad-Luke, Alastair. *Design Activism: Beautiful Strangeness for a Sustainable World*. New York: Routledge, 2009.

Groat, Linda and David Wang. *Architectural Research Methods*. New York: Wiley, 2013.

Hayden, Dolores. "The Power of Place: Claiming Urban Landscapes as People's History." *Journal of Urban History* 20, no. 4 (1994): 466–485.

Juhasz, Joseph. "The Place of the Social Sciences in Architectural Education." *Journal of Architectural Education* 34, no. 3 (1981): 2–7.

Kaplan, Rachel and Stephen Kaplan. "Bringing Out the Best in People: A Psychological Perspective." *Conservation Biology* 22, no. 4 (2008): 826–829.

Kavanaugh, Kelli B. *Detroit's Michigan Central Station (Images of America)*. Charleston: Arcadia Publishing, 2001.

Kopec, Dak. *Environmental Psychology for Design*. New York: Fairchild Publications, 2012.

Krinke, Rebecca, ed. *Contemporary Landscapes of Contemplation*. New York: Routledge, 2005.

Lepik, Andres. *Small Scale, Big Change*. New York: The Museum of Modern Art, 2010.

McGrath, Sean J. and Andrzej Wierciński, ed. *A Companion to Heidegger's Phenomenology of Religious Life*. Amsterdam and New York: Rodopi, 2010.

Mugerauer, Robert. *Interpreting Environments: Tradition, Deconstruction, Hermeneutics*. Austin: University of Texas Press, 1995.

Ocejo, Richard, ed. *Ethnography and the City: Readings on Doing Urban Fieldwork*. New York: Routledge, 2012.

Pilloton, Emily. *Design Revolution: 100 Products That Empower People*. New York: Metropolis Books, 2009.

Price, Travis. *The Mythic Modern: Architectural Expeditions into the Spirit of Place*. Novato: ORO editions, 2012.

Reason, Peter and Hilary Bradbury-Huan. *The Handbook of Action Research*. London: Sage Publications, 2006.

Ryan, Brent D. and Daniel Campo. "Autopia's End: The Decline and Fall of Detroit's Automotive Manufacturing Landscape." *Journal of Planning History* 12, no. 2 (2012): 95–132.

Seamon, David, ed. *Dwelling, Seeing, and Designing: Toward a Phenomenological Ecology*. Albany: State University of New York Press, 1993.

Seamon, David and Robert Mugerauer, ed. *Dwelling, Place, and Environment: Towards a Phenomenology of Person and World*. Boston: Kluwer Academic Publishers, 1985.

Smith, Andrew. *Events and Urban Regeneration*. New York: Routledge, 2012.

Sterrett, Joseph and Peter Thomas, ed. *Sacred Text, Sacred Space: Architectural, Spiritual, and Literary Convergences in England and Wales*. Boston: Brill, 2011.

Tilley, Christopher. *A Phenomenology of Landscape: Places, Paths, and Monuments*. Providence: Berg, 1994.

Tilley, Christopher. *Body and Image: Explorations in Landscape Phenomenology 2*. Walnut Creek: Left Coast Press, 2008.

Todres, Les. *Embodied Enquiry: Phenomenological Touchstones for Research, Psychotherapy, and Spirituality*. New York: Palgrave Macmillan, 2007.

Waldheim, Charles. *The Landscape Urbanism Reader*. New York: Princeton Architectural Press, 2006.

Index